THE CAUSES OF AGING

For Rosemary Elizabeth Boyce and Robin Kelly

CONTENTS

ABOUT THE AUTHOR

Photograph by Victor Trusch, Preston

Andy Wickens is Senior Lecturer in Psychology and Neuroscience at the University of Central Lancashire, UK. He obtained a BA (Hons) in Psychology in 1982 and went on to complete his doctorate at the University of Liverpool in 1988. He then spent two years working in the Pharmacology Department of Aberdeen University examining the pharmacological actions of cannabis-type compounds before taking up his present position in 1990. He has been involved in teaching Psychology and Biology for over 15 years which has also included tutoring for the Open University. His present research interests include neuropharmacology and the role of hormones, particularly testosterone, in mental function and aging. Drawing on his long experience of teaching, and broad knowledge of scientific research, Andy Wickens has the unique ability of being able to explain complicated scientific concepts in an easy and understandable way.

PREFACE

Growing old is like being increasingly penalised
for a crime you haven't committed

–Anthony Powell

Do not go gentle into that good night./
Rage, rage against the dying of the light

–Dylan Thomas

There is probably no other subject that occupies and fascinates the human mind as much as that of life and death. In life, aging is our constant and often unwanted companion that increasingly forces us to confront the unpleasant truth of our own mortality, and which brings the looming terror of our death ever nearer. It has been said that we spend about a quarter of our lives growing up and three quarters growing old – which is much the same as saying that we spend most of our lives in a state of irreversible decline. Along the way we are certain to have our youthfulness and vitality taken away from us, and be mugged by illness and disease. If we are fortunate we will see our friends fall by the wayside: if not then it will be our friends who will see us laid to rest. Make no mistake, despite what others may say, death is not a pretty picture. The noted gerontologist Bernard Strehler once remarked that he hated death. Yet death is not really the true villain. The real culprit that inflicts the greatest pain and suffering upon us is the process of aging that draws us ever onwards to death's door. Thus, if anything, we should vent our anger against aging – after all, death is only its dirty work.

Although it appears as if life and aging are inseparable, there are grounds for doubting whether they are as inextricably interwoven as they first seem. For example, it has been estimated that if we could maintain throughout life the same resistance to stress, injury, and disease that we had at the age of ten, then about one-half of us here today would be alive in 700 years' time. Thus by the time we reach adolescence, our bodies have been blessed with truly remarkable strength and vitality, and barring an accident or self-inflicted injury, we are in a state where we could essentially live for ever if only we could stop the aging process. Unfortunately, we all know that this cannot be done. But if we think about it more carefully, we can see that life has allowed us to be potentially immortal for a short period in our lives, and only then does something start to change, and we slowly lose what nature has generously given us. What then is the cause of the change that destroys us? Surprisingly, few of us ever seem to give the matter any serious thought. Yet, if we knew the answer, we would be in a much stronger position to do something about halting its progress. It was this realisation that first led me to find out more about the biological causes of aging – and ultimately to the writing of this book.

It still strikes me as odd that most people do not think more deeply about the reasons why they age and grow old. Perhaps it is because aging is so universal and inevitable that most of us simply take it for granted. Nothing in the universe is everlasting, and perhaps because of this we unconsciously assume that aging is as inevitable as the Earth spinning on its axis and the sun rising each morning. However, such a dismissive view is not totally justified. Obviously we can never hope to stop the Earth spinning, or orbiting the sun (nor would we want to!), but aging is different. For one thing it takes place within every one of us, and there are no cosmological forces at work that are beyond our control. In fact, whatever is responsible for causing aging must be at work in front of our very own eyes, or at least detectable and measurable with scientific instruments and methods. Thus, we are in a position to master the situation if only we could come to grips with its intricacies. Nor is aging something that is magical or unfathomable. In fact, it is clear that aging must have its *causes* like everything else in this world, and there is no good reason why science cannot discover *exactly* what these are. Obviously aging is complex, but this does not mean it is beyond our understanding or ingenuity. Thus, aging is not as inevitable as the sun rising each morning. Indeed, in a hundred years' time it is almost certain that the Earth will still be spinning on its axis, but the same cannot confidently be said about the existence of human aging as we know it.

This is a book which attempts to explain in simple and non-scientific language why we age and (in the last chapter) how this knowledge may one day possibly be used to control and overcome aging. In short, the book will focus mainly on trying to explain the fundamental and underlying *causes* of aging. I should say at the outset that nobody knows for sure what the true causes of aging are – although there is no shortage of theories. My intention is to guide the reader through the main ideas, and to present some of the most important evidence.

However, there are many difficulties in writing a book of this kind. For example, before one can understand the biology of aging, and concentrate on its underlying causes, one must, of course, have some knowledge of the body and the underlying biological systems of which it is composed. Thus, this book will cover some basic biology although hopefully it will be kept to a minimum. Another problem lies with the discipline of biological gerontology itself. This branch of science has grown enormously over recent years and to understand the subject in its entirety one has to be a specialist in all areas of biology. In fact, it is sometimes said that there are no such things as biological gerontologists – only biochemists, cell biologists, endocrinologists, immunologists and so on, all of whom are interested in aging. Indeed, by definition, one cannot be a specialist in all these areas, and I am no exception. I can only hope that the specialist who happens to pick up this book will grant me at least some latitude in my attempt at making the book accessible to a general readership. I make no apologies for trying to keep things simple.

As will be touched upon in this book, the history of aging research has a colourful and turbulent history replete with myth-makers, charlatans, quacks and broken reputations. Unfortunately the legacy of past controversies and misdemeanours still casts a long dark shadow over present-day aging research, and even now one can sense in the air that research in this field isn't quite respectable. Far better it seems to train people in understanding how the body works, or in conquering heart disease, cancer, senile dementia, and so on. Indeed, the biology of aging remains conspicuously absent from school curriculums, and there are relatively few courses that cover the subject even at university level. This neglect is curious since most people find the topic of great interest and personal relevance, and because one suspects that if aging was ever overcome then there would be few diseases left to worry about. Moreover, the situation continues to perpetuate itself. The future of aging research is clearly dependent on the quality and quantity of young scientists who are encouraged to enter the field – yet such people are never taught that such a subject exists! I like to think that this book can, in its own small way, help rectify the situation.

Although it is an unusual way of going about writing a book, I have tended to enlist the help of friends and students, rather than experts, to read the chapters and to provide me with feedback. Too many books, I feel, are not reviewed by the type of people they are aimed at, with the result that they often make perfect sense to fellow academics, but have little meaning for non-specialised readers. I hope my book does not fall into this category, and that it is comprehensible to everyone who is interested in the causes of aging. Amongst the people who have read parts of this book during its preparation, and provided helpful feedback are: Victoria Bretherton, Lynne Dawkins, Ian Ferguson, Roxanne Kelly and Lyndsay Wright. I would also like to thank Harwood Academic Publishers for setting up the project and also taking the trouble to review most of the chapters in their formative stages. Further, I am grateful to John Hodkinson for his help finding a suitable illustrator and to Mark Dickens for undertaking the artwork on the cover. And, finally I would also like to thank Andrew Hedgecock and Martina Deery for their time and effort, in proof-reading the chapters and attempting to correct the grammatical infidelities of my work. Despite all the good intentions of others, however, I stand alone as responsible for the book's inaccuracies, misconceptions and idiosyncrasies.

THE CHARACTERISTICS OF AGING

*Some people try to achieve immortality through their offspring
or their works. I prefer to achieve immortality by not dying*

–Woody Allen

*With a little luck, there's no reason why you can't live to be one
hundred. Once you've done that, you've got it made, because
very few people die over one hundred*

–George Burns

The forfeit that we pay for life is death. No human being in the history
of the world has ever avoided death and, unless there is some remarkable medical and scientific advance in the near future, nor will you.
This is not to say, however, that this situation will always exist.
Science has the potential to understand and solve the most intractable
of problems and this is likely to be true of aging. Because it appears to
be such an inevitable consequence of life, it is easy to lose sight of the
fact that aging is a biological process rooted in the laws of physics and
chemistry. Aging must have a cause, and if the cause (or causes) can be
identified then it follows that scientific and medical intervention into
the aging process may one day be possible. One can only guess at the
possibilities and implications; but the ultimate dream, of course, must
be the attainment of human immortality, and the banishment of illness and disease. The probability of this breakthrough occurring in the
immediate future is not great, simply because nobody, as yet, knows
for sure what causes aging. However, over the long term, it is almost
certain that aging will be conquered. Indeed, as I show in the final
chapter of this volume, it is entirely feasible that science will have discovered the means of stopping aging by the end of the next century,
although whether we choose to avail ourselves of such benefits will be
a completely different matter. But until that day arrives, aging is a fascinating biological puzzle that remains to be solved. This book is a journey through the landscape of the most important scientific theories

that are currently being used to explain why we age. There are no easy answers, but somewhere in this terrain lies the signpost that points to the place where the most complex and worthwhile biological problem of them all will one day give up its secrets.

WHAT EXACTLY IS AGING?

This deceptively simple question has no simple answer. In one sense we all know what aging is because we all live with its consequences and continually observe its effects in others. If pressed to define aging, we might reply that it has three main characteristics: (a) it is universal and inevitable, (b) it is gradual, progressing with time and resulting in an irreversible loss of fitness, and (c) it ultimately leads to death. This definition would probably conform to most people's ideas about aging. Yet from a strictly scientific point of view this definition confronts a number of difficulties.

The first problem is that not all living creatures grow old or age. For example, a somewhat surprising finding is that certain single-celled organisms such as bacteria and yeast do not age. These are simple creatures, admittedly, but this phenomenon raises the important point that aging is *not* an inevitable feature of all life. Therefore, our definition of aging only applies to certain types of animal. However, if aging is not a universal process this raises the rather awkward question of why aging should occur in the first place. This issue is discussed more thoroughly in chapter two; but suffice to say for now, it seems that the first forms of life to appear on Earth were immortal, and that at an early stage in evolution this strategy was abandoned for mortality. The question of why this change occurred has been the subject of much debate, although the suspicion exists that life once had (and maybe still has) the option of immortality but chooses otherwise. Therefore, aging is not universal, and in a theoretical sense it might not even be inevitable.

The question of inevitability can also be examined from the perspective of comparative life spans. For example, the shortest lived mammal the smoky shrew (*sorex fumeus*) lives barely a year whereas the human being can sometimes survive well over a hundred years. Both the smoky shrew and the human are the result of their evolutionary and genetic inheritance which must be the main determinant of their respective life spans. Obviously, it is inevitable that both are going to age at their own respective rates; but why has evolution acted to allow one to live for a few months and the other to exist for over a century?

When considering this question, it is hard to escape the conclusion that evolution has acted to program the length of life span in each species, rather than to leave it to chance. But if evolution has acted to program some forms of life to exist for a hundred years, why not two hundred years? Or even more? Again, the suspicion arises that evolution is potentially capable of making aging far from inevitable.

Our second defining feature of aging is the gradual decline of strength and physiological function (often called loss of vigour or vitality) that occurs with the passage of time. We all recognise this decline in ourselves and see it occurring in other forms of life. Research shows that practically every biological function in the body declines with age – our senses become less acute, our heart pumps less blood, our muscles become weaker, our mental abilities decline, and so on. There are probably hundreds, if not thousands, of examples we could use to illustrate the continual loss of function with aging. But not all biological functions decline at the same rate, and in some individuals (especially those on training or fitness programs) there may be some functions that *improve* with age. Thus, strictly speaking, decline need not be, and is not always, an inevitable consequence of aging. As we shall see later, we have a certain degree of control over the aging process and this indicates that aging may not be as predetermined as it first appears.

This is, perhaps, a good place to draw attention to the problems of defining the word *aging*. In its everyday usage 'aging' simply means growing older, but, it is important to realise that this is not the same as 'aging' in the biological sense which is defined as 'declining vigour'. It is generally accepted that a decline in vigour does not occur during development or growth, but only after maturity or adulthood has been reached. Consequently, a child or young adolescent does not 'age' in the biological sense. For this reason, most investigators prefer to use the term 'senescence' to describe aging that results from the decline in vigour following maturity (although this book for simplicity will continue to use the word 'aging'). Interestingly, the concept of senescence seems to imply that if only we could continue to grow (preferably very slowly) over our life span, then we would not age. This issue is discussed more thoroughly in chapter 11.

Finally, our definition of aging asserts that there is an increasing probability of dying with advancing age. This is undeniably true, and for obvious reasons, death is usually regarded as the end point of aging. However, we must always be careful to view the concept of death in its correct perspective. As we get older we become weaker and more susceptible to insults and diseases that are increasingly likely to cause death. Yet, despite this, it is probably the case that nobody ever dies of old age as such. Death is nearly always the result of a disease process, stress, or an accident, that causes organ failure. Rarely, if ever, does death occur because an organ simply gives up the ghost and packs up. Instead, we tend to die of heart disease, or cancer, and so on. Thus, aging is the underlying biological process that makes us more vulnerable to all of life's insults: it is not generally regarded as a disease, although clearly the two go hand in hand.

HUMAN LIFE EXPECTANCY

There is no doubt about it: we live in a fortunate age in terms of life expectancy. In ancient Greek and Roman times the average life expectancy was approximately 20 to 30 years; the Middle Ages boasted a life expectancy of around 35 years; and by the beginning of the 20th Century, this figure had not even reached 50 years. Today the life expectancy in the UK has risen to 74 years for men and 79.5 for women, and these figures are likely to rise further over the next few years. This recent increase in longevity is sometimes attributed to the advances in medicine and the improvement in living conditions that have occurred this century, although this belief is largely mistaken. The main reason why life expectancy has increased so dramatically is because of the decline in child mortality involving infants under one year of age. In 1901 the infant mortality rate was 150 per 1,000 or 15% of all live births. In other words, out of every hundred children born, fifteen would have failed to reach their first birthday. Today, the figure is just 7.4 per 1,000 (or 0.74%) and it is continuing to decline. Because of this improvement, the people who previously would have died during their first year are now tending to lead healthy lives well into old age. It is this increase in the number of early survivors that has done most to increase the average life expectancy this century.

This is readily illustrated if we compare the life expectancy of adults, rather than children, across different historical periods. For example, in Victorian England, around the time of the Industrial Revolution in 1841, a man who had managed to survive until the age of 65 years could expect, on average, to live a further 10.9 years of life. Today a male reaching 65 years has an expectation of 14.3 more years, which is only 3.4 years more than his Victorian counterpart. Although this difference is slowly increasing, it nevertheless remains the case that the apparent rise in life expectancy this century is largely an illusion. The main benefit of living in the 20th Century in terms of aging, it has to be said, is that we are simply more likely to survive past our first year.

We may be blessed with the best ever chance historically of reaching a grand old age, but it still remains true that for most of us, the average expectancy of 75 years does not seem particularly generous. Most would probably agree that the human life span is unfairly, if not tragically, short and in need of some upward revision providing it can be accompanied by reasonable state of health and fitness. Of course, a number of people will achieve this aim and succeed in living beyond the average span of 75 years with a considerable degree of vitality. But, for every winner there is a loser. In fact, nearly half of us will not reach our 75th birthday, and a substantial proportion of us will not even get close. To estimate our odds of reaching a certain age, statisticians construct what is called a life table which predicts how many people are likely to be alive after a given amount of time. Life insurance companies have long used life tables to work out the premiums they charge to their customers, but they also make fascinating reading for those of us with a vested interest in working-out our chances of reaching an old age.

At present in the United Kingdom there are approximately 800,000 new births each year with about 639,000 people dying to make way for them. This means that there are just over 2,000 new infants being born every day to inject new blood into the population. If we could somehow manage to take a thousand of these new births (with equal numbers of males and females) and follow them throughout life, what would we observe? According to recent statistics approximately 50% of men and 30% of women would fail to reach 75 years. The first hurdle is the first year of life (or to be more exact the first few weeks of life) and as we have already seen, a small percentage (0.7%) of infants will die during this period. Once an infant reaches its first birthday, however, future prospects become much more favourable. Around 8 infants will die by the age of 10, and by the age of 20 years the total number of deaths will probably have reached 13, or just over 1% of our sample. Over the next 20 years the number of deaths will slowly rise to a total of 28, with the result that there will still be over 97% of our cohort celebrating their 40th birthday. By the time our group reaches the age of 60, a mere 108 people (10.8%) will have died, and by the age of 70 years this figure will have risen to 260, or just over a quarter of our sample.

At this point, however, the number of deaths dramatically increases. Over the next ten years, around 450 people will draw their final breath, leaving behind less than a third of our sample by the age of

80 years. The vast majority of these remaining octogenarians will then die over the next decade, and practically everyone will be dead before they reach 100 years. At present only 9 people in every 100,000 (slightly less than 1 person in every 10,000) succeeds in reaching a hundred years of age. The prospect of hearing the letter box snap and the postman delivering our congratulatory telegram from the King or Queen (or President?) is not one that most of us can realistically expect to experience.

THE GOMPERTZ LAW

It is clear from the foregoing account that the risk of death increases with age, and that the majority (over 60% of us) will die between our 60th and 80th birthdays. However, there is more to life-span tables than first meets the eye. In fact, there are laws of aging at work that

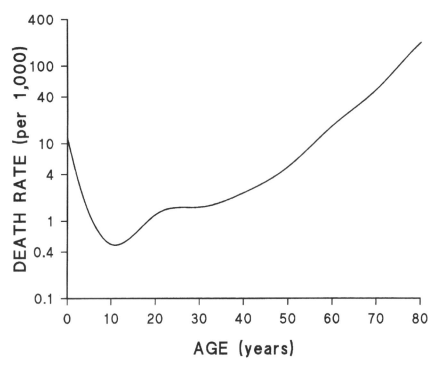

Figure 1.1 Gompertz mortality curve for a population characteristic of the developed world.

can be expressed mathematically. One of the first people to decipher the mathematical underpinning of the life-table was the English statistician Benjamin Gompertz, who in 1825 was employed as an insurance actuary responsible for working out life insurance premiums. These types of policy are essentially a gamble that the client will live (or not live) a predicted length of time, and consequently it was crucial that life insurance companies had reliable data with which to make informed and profitable decisions. Gompertz collected life tables for a number of regions including Northampton and Carlisle in England, and in doing so discovered a simple but fundamental law. In short, he found that the probability of death in the population doubled every 8 years after the age of 30. Graphically this can be expressed as a smooth straight line providing the axis recording the number of deaths (mortality) is scaled exponentially (2, 4, 16, etc.) (see figure 1.1).

In spite of all the advances of modern medicine, the relationship between age and death, first described by Gompertz, is as true today as it was over 150 years ago. In other words, it remains the case that after the age of 30 years, the rate of death in the population doubles every 8 years. That is, a person who is 38 years of age is twice as likely to die as someone who is 30, and someone who is 46 is twice as likely to die as someone of 38. Moreover, when Gompertz graphs are used to compare different human populations, a further remarkable finding is uncovered. Despite big differences in average life span across all types of human populations, the slopes of the Gompertz line actually turn out to be essentially the same (see figure 1.2). Regardless of whether one lives in an advanced industrial nation, or a third world country, the probability of death still doubles every 8 years or so after the age of 30. This finding is particularly surprising when one considers the different causes of death that occur between cultures. For example, in the USA the risk of breast cancer is ten times more likely than that found in Japan, yet the Gompertz graph for the two populations are almost identical. The reason for this consistency between widely differing populations with their own pattern of diseases is far from clear, but it points to one inescapable conclusion; there appears to be the same aging force at work wherever one lives, and whatever one's lifestyle. In other words, the Gompertz line appears to describe a general law of human aging.

Despite this, Gompertz's law does not apply to all age brackets. For example, the linear increase in mortality with age does not exist before the age of 30 years. Perhaps even more surprising is that the linear trend does not continue past the age of 90. In fact, beyond this age the slope of the Gompertz line for humans actually decreases and the

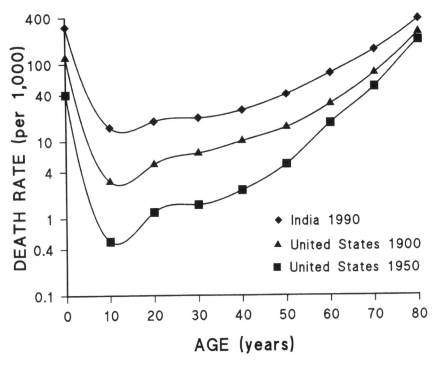

Figure 1.2 Gompertz mortality curves for differing developing countries (data redrawn – not to scale – from Jones, H.B. (1956) *Advances in Biological and Medical Physics*, 4, 281–337).

probability of death declines slightly. Recent evidence has lent support to this idea by showing that after the age of 105 years, the very small percentage of people remaining appear to have a greater chance of reaching 106, than individuals aged 104 have of reaching 105! It is almost as if exceptionally long-lived individuals are not adhering to the same laws of aging that apply to the vast majority of people.

Gompertz's discovery seems to provide a reasonably accurate mathematical means of describing human populations, but what about other animal species? This is a very important question: since if there is a general law of aging at work then we would expect it to apply to both humans and animals alike. Early studies following Gompertz's lead showed that most animals, including birds and small mammals, did not fit the Gompertz picture. But there was a problem with such a direct comparison because many animals live in harsh and dangerous environments where accidental death due to predation, lack of food, and so on, are common. Thus, most animals in the wild, unlike humans, do not survive long enough to show the signs of old age. In 1867, W.H. Makeham of the British Museum suggested a modification to Gompertz's formulae which attempted to take accidental death into account. Indeed, the inclusion of this new formulae showed that both human and animal mortality, despite their huge differences in life span, did fit with Gompertz's predictions. For example, the maximum life span of a mouse is about 4 years and the death rate doubles every 3 months. If a Gompertz graph is plotted for this data the slope of the line is found to be very similar to that found in humans. It appeared, therefore, that the same underlying beat of life that governed human mortality also extended to other animals.

Despite this, there has been recent growing concern over the validity of the Gompertz law to describe all types of species' life spans. One problem is that there is very little reliable data describing the death rate in large samples (or populations) of animals, and this makes it difficult to assess whether the Gompertz decline really does exists as a general law of aging in the same way as it does for humans. Indeed, in one remarkable study, James Carey and his colleagues started with a group of over 1.2 million Mediterranean fruit flies (*Ceratitus capitata*) that were all taken from pupae of the same age (these flies normally live between 3–4 weeks). As they counted the number of flies dying each day, they found (as expected) that the mortality rate increased in a constant fashion during their first 3 weeks of life. However, at this point, contrary to what the Gompertz law predicted, the mortality rate then levelled off and did not increase any further for the next 5 weeks. In fact, at around 8 weeks of age, when less than 0.1% of the original sample of were still alive, the mortality rate actually started to decline. As we have seen above, a declining mortality also seems to occur with very old humans, suggesting that there is something very special about the longest-lived members of any species – even flies.

The work with fruit flies shows that the pattern of aging decline may not necessarily be the same for all species thereby casting some doubt upon the universality of the Gompertz law. However, it must be remembered that insects are not mammals and it is not clear whether it is meaningful to compare the two in this way. Even if it is, it still remains the case that the majority of animals (even fruit flies) appear to show a Gompertz decline for at least part of their life. And of course, for humans, the evidence shows that the Gompertz model provides a very accurate fit for describing the rate of mortality up until very advanced ages. Thus, it does seem likely that humans (and *probably* most other animals) are subject to a fundamental law of aging that was first described by Gompertz. If this is indeed the case, then it must be that there is an underlying biological process at work to cause its effects.

WHAT CAUSES THE GOMPERTZ DECLINE IN HUMANS?

The Gompertz graph shows that our life span has two stages: a period of growth and maturity lasting up to about 30 years, and a much longer period of gradual decline. During the first stage, the young human body is fortified with extraordinary vitality and death defying properties, and indeed, the probability of dying from natural causes is at its lowest during our teenage years. In fact, Dr Alex Comfort has estimated in his book *The Biology of Senescence*, that if we could some-how maintain throughout life the same resistance to stress, injury and disease that we had at the age of 10, then about 50% of us could expect to be alive in 700 years time. This is mainly because the strength and capacity of the young human body, and all its various organs, is some 4 to 10 times greater than is necessary to sustain life. These findings make perfect sense from an evolutionary point of view. Death rates are at their lowest during the period of maximum reproductive capacity, whereas after the age of 30 (when presumably we have reproduced and passed on our genes) there is no further need to assure the survival of the individual. It is as if our bodies are specially designed to make absolutely certain we reach adulthood, but following this, our warranty runs out and we are left to start accumulating weaknesses that gradually increase our likelihood of dying.

The fact that an increased probability of death occurs as we get older indicates that we are getting weaker or less vigorous with age. But what exactly is the cause of this decline? One answer is that we are becoming increasingly more susceptible to disease. Gompertz provided evidence to support this theory by plotting the age changes in mortality for individual diseases such as heart disease, cancer, stroke, and pneumonia, etc., and found that the death rates for each of these diseases followed a straight line that closely paralleled the general Gompertz line (see figure 1.3). Thus, in one sense, the increasing incidence of diseases can explain the aging decline. However, this explanation is not entirely satisfactory because aging and disease can be shown to be totally different things. For example, although disease tends to accompany aging, most researchers agree that we can never-theless age *without* disease. Thus, disease is not the main cause of the

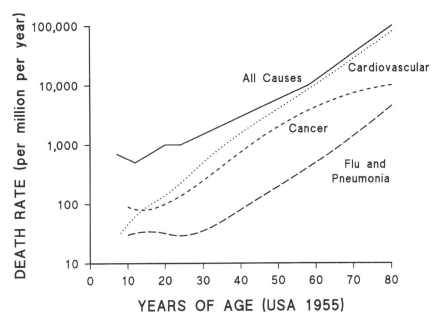

Figure 1.3 Mortality rates across the life span for various diseases *circa* 1955 (data redrawn – not to scale – from Harrison, D.E. (1982) *Biology Digest*, 8, 11–25).

Gompertz decline, and to properly explain the increase in disease one must try to understand the underlying aging process.

Aging has many characteristics but the most important from a biological point of view is the gradual decrease of strength and function that occurs in the organ systems throughout the body. Whatever system one decides to examine, one tends to find age-related decline. For example, the average amount of blood pumped out from the heart decreases from around 5 litres per minute at the age of 20 to about 3.5 litres per minute by the age of 75, representing a decline of about 0.7% per year. The same trend occurs for maximal heart rate which decreases from about 220 beats per minute in the young adult to about 160 beats at around 65 years. According to Sherwin Nuland in his book *How we Die*, maximal heart rate falls by one beat per year, a figure so reliable that we can work out our rate by subtracting one's age from 220. The lungs also show a gradual decline with maximum breathing capacity decreasing from about 165 litres per minute at 25 years to about 75 litres at 85 years. The kidneys show a 50% decline in filtration rate over the same time span. Other declining functions include decreases in muscle mass, hormone levels, visual acuity and the size of the brain (which falls from about 1,375 grams at the age of 30 to 1,232 grams at age 75). It should be remembered, however, that these are average figures and that considerable variation exists between individuals.

What then is the relationship between the decline in biological function and the Gompertz graph? The answer is that the two appear to be intimately related, although the extent to which decline in an organ system parallels the Gompertz slope depends on the type of system being examined. Much of the work examining this issue has been undertaken by Nathan Shock who founded the Baltimore Longitudinal Study of Aging in 1958. This is a ongoing program of research that measures the performance of individuals on biological and psychological tests across their life span. Subjects voluntarily give themselves up

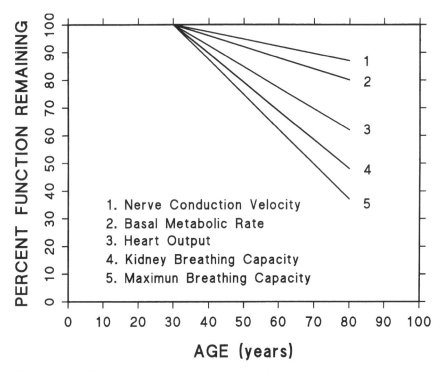

Figure 1.4 The decline of function for physiological systems of varying complexity across the life span (data redrawn – not to scale – from Fries, J.F. and Crapo, L.M. (1981) *Vitality and Aging*: W.H. Freeman. Original data from Strehler, B.L. *et al* (1960) in The Biology of Aging: A Symposium: American Institute of Biological Sciences.

for 2 days of extensive testing every 18–24 months thereby allowing their age-related decrements in performance to be accurately measured. This work has shown that the functions which show the most severe decline with age are those that require the co-ordinated activity of a number of organ systems. For example, the speed of nerve impulses changes by only 10% between the ages of 20 and 90, whereas maximum breathing capacity decreases by 50%. The former is simply a measure of a single nerve cell, whereas the latter depends on the co-ordination of a large numbers of nerves in various nervous systems, along with combined activity in muscle and circulatory systems (see figure 1.4).

These, and a large number of other findings, show that the various systems and functions of the body age at different rates. Furthermore, this is just as true of mental performance. For example, some simple measures of performance such as reaction time or vigilance may show little decline in healthy individuals up to about 70 years, whereas other more complex mental tasks may start to show an appreciable decline by middle age. Thus, whether we examine biological or mental performance, aging appears to exact its price by impairing maximum or complex performance. Unfortunately, life itself is the most complex system of all, requiring the integration and co-ordination of hundreds, if not thousands of different biological systems. The relationship between age and mortality which Gompertz first described in 1825 would therefore appear to be the sum accumulation of all these decrements, both simple and complex.

A CAUTIOUS NOTE

There is little doubt that we are able to age and grow old without the help of disease. However, this does not mean that disease is unable to influence aging. A recent study from the Baltimore Longitudinal Study of Aging examined a group of males, aged 65 to 80 years, who were rigorously screened to exclude heart disease. Remarkably, these disease-free subjects were found on average to have the same cardiac output (the amount of blood ejected by the left ventricle) in response to physical exertion as that found in younger men aged 25 years. Although the older subjects showed a reduction in heart rate with exercise (this is a reliable feature of aging as noted above), they nevertheless compensated for this deficit through a greater increase in the stroke volume of their blood. In short, these findings show that disease may contribute to some of the bodily changes that are commonly viewed as aging, as well as showing that aging does not lead to a decrement in all measures of heart function.

These findings are important since autopsy surveys show that approximately 60% of men who live to their sixth decade have a 75–100% narrowing in at least one of their coronary arteries. Consequently, many previous studies that have attempted to describe the effects of age on the cardiovascular system may not have studied age as such, but aging as caused by coronary heart disease. Moreover, heart function does not provide the only example. For example, research undertaken at the National Institute of Aging examined the utilisation of oxygen by the brain which had previously been believed to decline with aging. However, when a group of subjects (aged between 20 and 80 years) were carefully screened to be free of brain disease (by their performance on a battery of neuropsychological tests), it was found that oxygen metabolism remained unaltered with advancing age.

These findings show that as researchers study older people more carefully, some of the changes that in the past have been called 'normal' aging are in fact due to disease processes. Thus, the independence of aging and disease is not as clear cut as it first appears. Although most investigators accept that aging and disease are not the same, it may be the case that subclinical disease is a more important contributor to physiological decline (and what *appears* to be aging) than first meets the eye. These findings raise many more questions than answers, and show just how little we know about the aging of body function. Despite this, it remains the case that physiological decline is the main feature of aging, and still provides the best explanation of the Gompertz decline. Moreover, if there is a lesson to be learnt from this work it is this: if you want to minimise your chances of aging, stay healthy.

HOMEOSTASIS AND AGING

There is much more to biological decline, however, than simply a decrement in strength and maximum performance. It is easy to take life for granted, but it depends crucially upon the internal chemistry of the body working within very fine limits. For example, our body temperature has to remain within a few degrees of 37°C (98.6°F) for us to remain alive. If it increases to 43°C then fatal haemorrhaging will result, and if it falls below 26.4°C heart failure will occur. Similarly, the glucose concentration in our blood is usually maintained at 90 mg per 100 ml of blood, and this tends not to fluctuate (except after a large meal) by more than 0.1%. Some types of body constituents have to be kept within even more strict limits. For example, a small rise in the amount of potassium in the blood will quickly stop the heart, and a slight increase in magnesium will block nerve function. Thus, for a person to stay alive, the body must be able to constantly maintain a precise set of conditions in all of its chemical and physiological states. To ensure this, it is crucial that each variable must not be allowed to stray above or below an acceptable range.

The body is constantly striving to maintain its internal constancy despite being frequently subjected to considerable stress and change. For example, a simple activity such as walking will raise body temperature, burn nutrients for energy, and cause a loss of water from sweating and breathing, all of which could lead to serious repercussions if they were not carefully monitored and controlled. The process by which the body manages this feat is called homeostasis and this important mechanism requires a complex set of interacting systems to be in place. To start with there must be *detectors* in the body that are able to sense the changes taking place. This information must then be relayed to a *control centre* that determines the appropriate response with respect to the demands of the situation. And finally this information needs to passed on to the *effectors* that set in motion the adaptive response. Thus homeostasis requires a system of at least three interacting components and these can be illustrated by the example of temperature regulation. For instance, when body temperature rises above a certain point, the hypothalamus (a small structure in the brain) detects the change, decides upon the required response and then sends neural signals to the sweat glands (the effectors) to increase perspiration. If temperature drops, then a new response leading to shivering may be initiated. Most of the time, however, the process of temperature homeostasis ticks away without having to go to either of these extremes. Indeed, without homeostasis our life support systems would run riot and life cease to exist.

One of the most characteristic features of growing old is the marked decline that occurs in homeostatic efficiency. Temperature regulation, again, provides a good illustration. As we have already seen, in humans it is crucial that temperature is maintained at about 37°C. In fact, under resting conditions this value changes little across the life span so that an old person has much the same body temperature as a young person. However, when greater demands are placed on the system a different picture emerges. For example, in a study performed in 1950, naked subjects were placed into cold cabinets for up to 2 hours and exposed to a temperatures of 5–15°C. The results showed that old subjects ranging from 57 to 91 years were less able to maintain their body temperature compared to a group of young subjects who were aged between 22 and 36. In fact, most of the old subjects experienced a drop in temperature of more than 1°C, and were less able to invoke a vigorous shivering response. These findings also help to explain why death from hypothermia significantly increases in the aged.

Similar findings are also found with heart rate and exercise. For example, during strenuous exercise young men may increase their heart rate up to over 200 beats per second, whereas older men on average may only manage to reach 160 beats per minute. Not only do older men take longer to reach their peak response, but they also show a much slower return to normal functioning of heart rate, oxygen consumption, and carbon dioxide elimination when exercise is finished. In fact, the slower rate of recovery with increased age is perhaps the most characteristic homeostatic response of all. To give but one further example: normally the glucose level in blood shows very little change with age, but if glucose is injected into the blood (mimicking the body state following a large meal) then the rate of restoration back to the normal level tends to be much slower in older people. This delayed glucose response may not be entirely trivial since, as we shall see in chapter 5, there is evidence that it may cause damage to protein and other body components.

It is impossible to describe all the age-related changes that occur in man, but it is clear that aging results in an impairment of homeostasis, along with a decreasing ability to achieve maximum physical (or mental) performance. In short, it reduces the ability to adapt to the traumas of life and this is undoubtedly one of the reasons why old people are much more susceptible to life threatening events. For example, a bout of influenza, a heavy fall, or forced physical exertion are likely to have no harmful effect on a young person but may well be sufficient to kill a frail old person. It is clear, therefore, that the decline in homeostasis and peak physical performance contributes significantly to the increasing trend in mortality as first shown by Gompertz.

JAMES FRIES AND THE RECTANGULAR CURVE

One of the most fundamental questions concerning aging is whether there is an absolute biological limit to the human life span. As we have seen, there is considerable physiological and homeostatic decline with aging, but how long can the body continue to function given an ideal set of conditions? This is a difficult question to answer, but one way of tackling it is to ask: what would happen if we could somehow banish all disease from the aging process? Without disease we would presumably live longer, but how long? In 1973 the National Center for Health Statistics in the USA attempted to address this question by working out the consequences for human longevity if all of the major causes of death were eliminated. The results of their study showed that if cardiovascular disease was totally eliminated, the average life span would increase by 17.5 years. If all of the most common types of cancers were eradicated, then a further 2.5 years could be enjoyed. And if murders, suicide, accidents, cirrhosis of the liver, influenza and diabetes were also abolished, a further 2.5 years could be added to the overall total. In sum, if all the main causes of death were eliminated we would gain 22.5 extra years. In 1973 this meant that the average man would live to be 92.5 years and the average women to 97.5 years. The underlying implication of this work is that the average natural human life span would appear to be fixed at about 95 years.

Another advocate of the idea that there exists a fixed life span, beyond which we can not go, is the Stamford rheumatologist James Fries who believes that the human body is biologically destined to succumb to death at around the age of 85 (give or take a few years). Fries arrived at his conclusion by plotting how life expectancy from birth has changed in the past century, and then combining this with a plot of present-day life expectancy from the age 65 (see figure 1.5). It can be seen from the graph that the two lines converge at about the age of 85 years. Fries reasoned that since life expectancy at birth can never exceed the life expectancy at 65 years, then the intercepted lines must reflect the upper average age limit of human life span (although some researchers have pointed out that this is a rather strange form of logic!). Of course, some people will die well before this age and some will survive longer, but according to Fries, 85 years is the average maximum age we can hope to live to.

Fries' work also leads to a number of important predictions concerning the way our population is likely to age over the next few decades. To illustrate his predictions, Fries uses what are known as population survival curves. These graphs are not unlike those used by Gompertz, except that instead of plotting the death rate at different ages on the vertical axis, Fries transforms this data into the cumulative number of people surviving at different ages (see figure 1.6). This simple change produces a very different type of curve that tends to be rectangular in shape, and one which casts new light on how life span is likely to change in the near future.

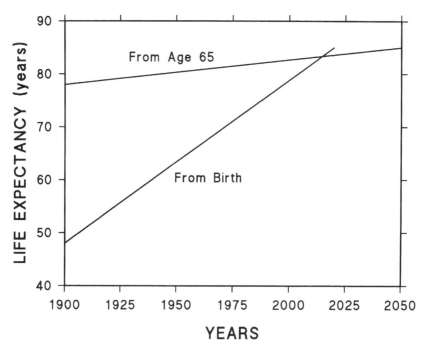

Figure 1.5 Graph showing the interception of life expectancy at birth with life expectancy at 65 years (data redrawn – not to scale – from Barinaga, M. (1991) *Science*, 254, 936–938).

If survival curves are constructed for different historical ages (or even different societies from third world to advanced) it is found that the curves become more rectangular in shape. The upper horizontal line or top edge becomes straighter and longer, showing that fewer people are dying at early ages. The down line on the right side of the rectangle, in contrast, signals the time when most of the population will start to die. Interestingly, as this line has become straighter and more vertical, it shows that the majority of deaths will occur in a much shorter time frame. If this trend is indeed true, and if Fries is correct in arguing that the average length of life is fixed at 85 years, then it has profound implications. In short, it shows that as the down slope of the rectangle becomes progressively more vertical, the period in which most people can expect to die will get condensed. Or, as Fries puts it, there will be a "compression of morbidity".

This is actually very good news. Increased rectangularization of the survival curve indicates that most people will enjoy longer, healthier

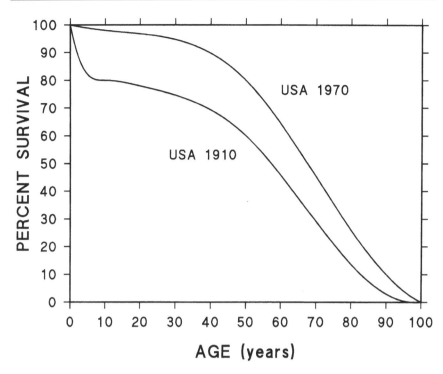

Figure 1.6 Example of rectangular curves from two different periods in the USA (data redrawn – not to scale – from Fries, J.F. and Crapo, L.M. (1981) *Vitality and Aging*. W.H. Freeman and Co.)

lives, and that the onset of life threatening disease will tend to occur later on in life. Of course, mortality cannot be postponed for ever, and when the terminal decline does begin the graph shows that it will tend to occur suddenly and with devastating effect. Fries believes that even if disease has not played a part in contributing to death, the body will nevertheless fall apart like "the bursting of billions of bubbles all at once". In other words, people will simply die a natural death. Even so, this period of terminal decline should be mercifully short and the over-all net result will have been a long life full of vigour and vitality, ending suddenly and predictably.

There are, of course, also broader social implications of the rectangular curve. For example, if Fries is correct then the common stereotypical view of the elderly becoming frail and a burden to society may be wrong. Instead, Fries predicts that as the population rectangularizes the elderly will be physically, emotionally, and intellectually healthy and vigorous until shortly just before their death. This view has some surprising implications for the planning of future health resources because it predicts that chronic health disorders will come to occupy a smaller percentage of the life span, and consequently the needs for medical care of the elderly will actually go down. Needless to say this view is in stark contrast to nearly all other predictions regarding future medical care of the elderly.

WHAT IS THE MAXIMUM LIFE SPAN?

Fries' arguments are cogent and provocative, but they rest on a number of assumptions that are open to question. His most controversial assertion is that the average life span is fixed at about 85 years. As we have seen, Fries has obtained this figure by plotting how life expectancy at birth has increased over time, with life expectancy at 65 years, and then projecting these lines forward to see where they cross. Although Fries' belief that life expectancy at birth can never exceed life expectancy for those aged 65 sounds plausible, other researchers are not so sure of his logic. For example, Fries' theory only works if life expectancy increases in a straight line, and if either of the two life expectancy lines change from their linear course, the theory comes unstuck. As we shall see later, there is some tentative evidence that indicates that the shape of the lines are indeed beginning to change. Furthermore, Jay Olshansky of the University of Chicago has also pointed out that the data used by Fries in his graph includes infant mortality rates, and other types of death, that have nothing to do with the biological limits to life. Thus, it is not absolutely clear whether Fries' graph is as meaningful as it first seems.

But perhaps most controversial of all is Fries' insistence that there is a fixed limit to human life span. Indeed, without this anchor Fries' theory would be invalidated as his rectangular curve would never become a true rectangle because its bottom right corner would always be moving to the right. To support his claim, Fries points out that the *maximum* human life span has changed remarkably little since recorded history, and that the oldest people in former times lived to ages that were not greatly different to those of today. For example, the ancient Egyptians described the maximum life span as 110 years, and examination of tomb inscriptions from Roman times shows that a small number of people (about 3%) lived beyond 100 years. Although these figures can not be proven, and some are undoubtedly exaggerated, it is nevertheless clear that a few individuals did live well into old age. For example, the Greek dramatist Sophocles wrote Oedipus Rex at the age of 75 years, and won a prize for literature when he was 85. It is also recorded that the renowned Roman orator Marcus Seneca remained active and lived until he was 93. Thus, in spite of the marked increase in *average* life expectancy that has occurred throughout history, the *maximum* length of life seems to have remained relatively unchanged.

A more objective way of addressing this issue was made possible by the introduction of compulsory birth certification which began in England in 1837, and become established in most other industrial countries soon after. These records have provided researchers with a large and reliable pool of data that has enabled them to record accurately the changes in mortality that have taken place over the last 150 years or so. As we know, average life expectancy has clearly increased, but the important question by which to judge Fries is whether the average *maximum* life expectancy (which Fries puts at 85) is constant or not. Unfortunately this is difficult to judge because life expectancy is still less than 85 years, and while it is on the increase, there is still some way to go before the target is reached. Therefore, one has to use other methods to answer the question. One approach is to count the number of people reaching 100 years and beyond. Although Fries claimed in 1980 that there has been no increase in the number of centenarians in England and Wales since birth records began, this is not supported by recent statistical evidence. For example, there were a total of 2,280 living centenarians in 1981 and 4,390 in 1991; that is an increase of 93%, and this figure is likely to treble by 2001. These dramatic increases cannot be explained by any increase in birth rates between 1881 and 1891 since the number of births rose only marginally during this period. Thus, a far greater number of people are now living to 100 years. These findings are difficult (although not necessarily impossible) to reconcile with the assertion that average life expectancy is fixed at 85, and the implication is that Fries' estimate may have to be revised upwards. Perhaps only time will tell whether average maximum longevity is really fixed at 85 years or not.

WHAT IS THE GREATEST AGE ATTAINED BY A HUMAN BEING?

This question fascinates all of us. Although there will always be claimants to the title whose age cannot be proven, there is no doubt that the advent of birth certification has allowed an accurate assessment of this question to be made. Even without birth records it is possible to make mathematical predictions, and back in 1825 Gompertz estimated from his life tables that the greatest age obtainable was in the region of 110 years. This estimate has turned out to be remarkably accurate, or at least it has until fairly recently. In 1980 the Guinness Book of Records listed only 5 people in the world who had been reliably documented as living beyond the age of 112 years, with the oldest being Shigechiyo Izumi of Japan at 114 years (Izumi died in 1986 reaching 120 years and 237 days). This record has since been broken, by a French woman, Jeanne Calment who died on 4th August 1997 aged 122 years, five months and 14 days. She was born on February 21, 1875, outlived 17 presidents of France, and remembers selling crayons to Vincent Van Gough ("He was not nice. He was ugly with this horrible shepherd's hat and he smelt of alcohol"). Amongst the things that Mme Calment attributed her longevity to were: "destiny", regular glasses of port, a diet rich in olive oil and the ability to "keep smiling" (as perhaps best typified by her remark "I've only one wrinkle – and I'm sitting on it!"). There are now at least a dozen people who have made it beyond 112 years, and this figure will probably double over the next ten years. Again, this points to the possibility that maximal life span may well be inching slowly forward.

Of course, there have been many other people who have claimed to be older including Shiali Mislimov of Azerbaijan who died in 1973 at the reputed age of 168 years, and Charlie Smith of the USA who was reputed to have been 137 years in 1979. But in all these cases, as in many others, there is no firm documented evidence to support their claims. Indeed, there may be many benefits of exaggerating one's age, including the prestige and honour which is likely to result. Perhaps the best reason of all to be sceptical about such claims is the sheer number of people whose birth and death have now been recorded. Nearly all developed countries have kept accurate records for over 150 years, and yet only a handful of people out of countless millions have ever made it past 112 years. And this is despite people living in societies where there is ample food and advanced medical care. This is surely the most convincing proof that reported ages of over 120 years in places where no reliable records are kept must be viewed with considerable scepticism.

NATURAL DEATH

One of the most debatable aspects of Fries' theory is his belief in "natural death". According to Fries, as people live longer and avoid disease, the decline in vitality and function that accompanies aging will eventually lead to a natural breakdown of the body's resources and death. One gets the impression when reading Fries' work that it is almost as if the human body self destructs as it reaches the end of its life. This idea is not one, however, that is generally shared by the medical establishment which tends to view death as resulting ultimately from disease or accidental events. Although Fries may have overstated his case, there is nevertheless some evidence which shows that as people get older there is an increasing likelihood that they will die from a death with no clearly defined cause. For example, in a study undertaken by Robert Kohn it was found that no clear cause of death could be determined in 30% of autopsies undertaken on people aged over 85. However, to call such deaths 'natural' in the sense that they have no 'cause' is somewhat misleading. What is much more likely is that no cause of death can be identified that is of the magnitude that would cause death in a younger person. For example, minor chest infections or gastric problems may be the final cause of death in a very old person, but these would have little effect on someone who was younger. Perhaps these findings lend better support to the idea that ever smaller challenges to homeostasis will eventually have fatal consequences. Thus, if the concept of natural death is to be useful, it needs to be viewed as resulting from the declining capacity of an aging organism to withstand any given insult. Death it seems cannot occur independently of at least some form of trauma.

With increasing numbers of people living to advanced ages there has been a growing awareness from post-mortem research that the causes of death are showing some surprising trends in the very old. For example, an Icelandic study which looked at the mortality resulting from cardiovascular disease showed that the percentage of deaths was much higher in people aged 60–70 than for a group of people aged over 90 years. There is also some similar evidence from cancer mortality that shows the older the group the smaller the proportion of deaths from malignancies. Thus, not all diseases show a linear increase with aging. However, other types of death do show a marked increase in the very old. For example, a British study has indicated that pneumonia accounts for between 30–40% of deaths in people aged 85 and above. Furthermore, there is also an increasing likelihood of gastric and duodenal ulcers, as well as types of trauma (particularly from fractures of the hip) with increased aging.

CAN WE AVOID THE AGING DECLINE?

The theme throughout this chapter is that aging produces an inevitable decline in vigour which, in turn, increases the probability of disease and death. Although this statement is undoubtedly true when applied to large groups of subjects, it is less valid when we examine aging on an individual basis. This is not to say that aging can be stopped, but it appears that age-related change can often be slowed down, and in some cases even reversed. Strange as it may seem, we may have a considerable degree of control over the unfolding of the aging process.

To appreciate how aging can be controlled it is necessary to understand the concept of variation. Apart from identical twins, no two people are born exactly the same because of their genetic inheritance. The variation in genetic makeup and inheritance means that individuals will develop (with the help of environmental influences) different levels of biological performance, temperament and mental abilities. From a purely statistical point of view it might be predicted that the difference between individuals on any task will be greatest when their performance levels are at their highest before the age of 30. In other words, we might expect the difference in performance between different individuals to get smaller as they age. But, surprisingly, the reverse is true. What happens is that performance between individuals actually shows greater variability as they get older. That is, the difference between the strongest and the weakest is much more pronounced in groups of old people than in groups of young.

One reason why this occurs is because the aging decline is not as fixed or inevitable as the Gompertz or survival curves would have us first believe. To the contrary, in many cases physiological performance can be modified and improved. James Fries has neatly illustrated this using the example of marathon running. If the world record times for men aged from 10 to 79 are examined, then one can clearly see how the line of maximum human performance follows the path we come to expect of aging – namely that performance is maximal from the ages of 20 to 30, and then declines in a predictable fashion throughout the life span (Figure 1.7). This decline occurs at the average rate of approximately 2 minutes per year. But, if we turn our focus away from the world record holders, and examine the times of normal marathon runners, a completely different picture emerges. If we take at random a large group of 40 year old runners, we may well find that the time taken to run a marathon varies from 150 minutes to over 24 hours. However, the actual times are not really important. What is more

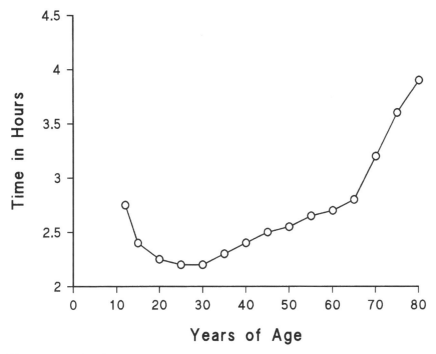

Figure 1.7 World record times for marathon running for different ages (data redrawn – not to scale – from Fries, J.F. and Crapo, L.M. (1981) *Vitality and Aging*. W.H. Freeman. Original data from *Runner's World*, 1980).

crucial is the fact that most of this sample, if not all, will be capable of improvement with practice and training. Increasing age is not a barrier to improving performance – except possibly in the case of world record holders. Bearing in mind that declining vigour is supposedly such a vital characteristic feature of aging, then this potential for increased performance is clearly a bonus for those determined to slow down their aging decline.

Running may be a simple activity, but it also seems to have many benefits for slowing down other aspects of aging. For example, it has been shown that joggers typically lose weight, have reduced blood pressure and cholesterol, and enjoy improved maximum breathing capacity. All of these measures tend to show an inevitable and linear decline with age in the population as a whole, but on an individual basis they are open to a certain degree of improvement. Psychological benefits, including increased self esteem, may also occur from jogging. Thus, running has the potential to benefit many aspects of life. This is not to say that everybody should take up jogging. Any type of exercise, or hobby, if taken seriously enough, will probably result in some benefits in the battle against aging.

The benefits of exercise have also been shown by the psychologist Waneen Spirduso who measured the speed it took to respond to a visual stimulus presented on a computer screen, in individuals who regularly played racquet sports such as squash. The results showed that sport-playing subjects who were aged between 50 and 70 had reaction times that were very similar to those found in less active people aged 20–30. Aged subjects who did not play these sports, however, had much slower responses. Since reaction time can also be used as a way of measuring nerve conduction velocity, one can therefore assume that the older sport playing subjects were probably more efficient on a wide range of performance measures, and in one sense 'less aged'. Regular exercise would therefore appear to be at least one way of helping to slow down the Gompertz decline.

Perhaps the most important anti-aging protection of all is having a positive psychological outlook. This was shown in a study (reported in Fries and Crapo's book *Vitality and Aging*) that looked at the effects of increasing personal choice and responsibility in nursing home residents. Two groups of old residents were compared, both of whom were given essentially identical care, except that one group were encouraged to be independent and make their own personal decisions. After only 18 months of the study, the results showed that 15% of the 'independent' group had died, compared to 30% of the 'cared-for' group. Responsibility and personal freedom would also appear to be crucial factors in the aging equation along with fitness and exercise (perhaps no better example of this is Jeanne Calment who took up fencing at the age of 85, and was still riding a bicycle at 100).

INTRODUCTION TO BIOLOGICAL THEORIES OF AGING

As we have seen, aging results from a declining loss of vigour and/ or adaptability that makes us more susceptible to disease, or death, as we grow older. Yet, so far, we have avoided the important question, namely, what are the underlying biological changes that cause this decline in the first place? This is the crucial question which most of the remaining chapters of this book will try to answer. The quest is one fraught with considerable difficulty, not least because of the great complexity of life, and the many different levels of explanation that are possible in describing a biological process. For example, at the molecular level there are the genes that are to be found in practically every cell in the body, along with the many complex biochemical processes that they initiate and mediate. This is an entire biological world within itself, and it would perhaps be surprising if these processes were not in some way involved in causing aging (although it is within the realms of possibility that they might not). The next level of biological organisation is the smallest unit of life, namely the individual cell, which is home to the genes and all their biochemical reactions. The human body contains trillions of cells that form the various organs and structures of the body. Not surprisingly, cell loss and the decline in cellular function have also been linked to the aging process – but again, despite being highly plausible, there is still considerable room for doubt whether this is the level where the causes of aging begin. Both the genetic and cellular theories can be called *intrinsic* explanations since they see aging as occurring from within the cell. However, at a higher level still, there are the *extrinsic* interacting organ systems of the body, along with neural and hormone mechanisms that provide the communication between them. In fact, it is possible to argue that aging is the result of events happening at this higher level without resorting to genetic and cellular explanations. But where do we start, and at what levels do the aging changes first begin? This is just one of the many mysteries of aging.

To make the problem even more difficult, changes that take place at one level are also likely to work their way up (or down), and influence other levels of biological organisation. It can be difficult to visualise how genetic changes, for example, can affect endocrine function (or vice versa), but such inter-relationships exist, and are potentially important in understanding aging. Linked to this problem is the difficulty in trying to establish *cause* from *effect*. No biological system exists in isolation, and a change in one system will tend to produce changes in other systems of the body. But trying to determine what biological change came first and what are the secondary manifestations of this change is quite another matter. It has been said that there is no other area of biological inquiry that is underpinned by so many theories as the science of gerontology, and this is perhaps not surprising when one considers the great complexity of the puzzle, and the fact that researchers do not even know for sure where to start looking for the causes of the problem.

The lack of understanding goes even deeper still. At the heart of the debate of what causes aging is the issue of whether aging is a predetermined process that is programed to occur (in the same way as the onset of puberty or the menopause), or whether aging is due to the random accumulation of damage which eventually reaches a level that is incompatible with life (i.e., wear and tear). The first view implies that we are genetically determined in some way to die, and the second implies that we simply wear-out. Indeed, this is possibly the most fundamental question of all in aging research, and without an answer to this problem, aging will always remain a mystery. This issue along with many other questions and uncertainties will confront us as we go through this book, but hopefully, as I will show, science has the problem of aging very firmly set in its sights, and it is surely only a matter of time before its intricacies are fully or partially solved. And, as I hope to demonstrate in the final chapter, the rewards of this knowledge are likely to be far reaching and maybe beyond our wildest imagination.

SUMMARY

- The decline of vigour that starts to occur after early adulthood is called senescence although the term *aging* is adopted in this book.
- While human life expectancy has been increasing throughout history (over the last century this has been largely due to the decline in infant mortality) maximum life span appears to be unchanged.
- The Gompertz law shows that the probability of death for human beings doubles every 8 years after the age of 30 years. This relationship, however, breaks down for the oldest of the old.
- Other animals also appear to have their own characteristic Gompertz decline, at least for part of their life span, although more work needs to be undertaken in this area.
- The Gompertz decline is largely due to the decline in physiological function that occurs with aging, although disease may also contribute to the aging decline.
- One of the best ways to understand aging is to view it as a progressive decline in homeostasis and reduced adaptability to stress.
- According to James Fries there is a maximal limit to life span (which he believes to be around the age of 85 years) and consequently as people live longer due to better health, the period in which people die will become condensed (that is, there will be a "compression of morbidity").
- The oldest person who has ever lived is probably Jeanne Calment who died at the age of 122 years, 5 months and 14 days.

- Our aging decline is probably not as fixed or inevitable as it may first appear. Training programs, exercise or just continued use of a function may help retard (or even reverse) certain parameters of aging.
- Complex biological systems have different levels of biological organisation (genes, cells, organs, etc.,) and it is clear that aging affects all levels. However, it is not clear whether the causes of aging spring from one level, or from all levels. The problem is confounded by the fact that nobody knows whether aging is a programed process or whether it is due to wear and tear.

THE EVOLUTION OF AGING

*Discovery consists of seeing what everybody else has seen
and thinking what nobody has thought*

–Albert Szent-Gyorgyi

Nothing in biology makes sense except in the light of evolution

–Theodosius Dobzhansky

There are only two fundamental questions we need to ask about aging: why it occurs and how? The question of *why* living creatures grow old and die has fascinated philosophers and writers for thousands of years. But in more recent times the issue has come under increasing scientific scrutiny, particularly when viewed against the backdrop of Darwin's theory of evolution which has transformed the way we understand life. As we shall see, life span has evolved like any other biological characteristic and it can therefore be assumed to serve some type of specific *purpose*. Moreover, understanding how and why this has occurred is likely to provide considerable insight into the causes of aging. In fact, if we do not fully appreciate the reasons why aging exists in the first place, and the mechanisms by which nature has enabled it to happen, then the chances are that we will never fully grasp its true secrets. But does aging really have a purpose? A physicist might point out that everything in nature has a finite existence, whether it is our Earth, solar system, or even our universe. Nothing is forever, and perhaps life is just a smaller scale version of this process. The physicist's view is probably correct to some extent, but it nevertheless remains a fact that the aging of a biological system is a very different thing to the 'aging' of inanimate matter (such as a corpse). Life follows different rules. It has, after all, raised itself above the physical world, and therefore it is perhaps more meaningful to view aging as the gradual weakening of life's own forces rather than passive physical decline.

WHAT IS LIFE?

Life defies simple definition, but its main features are simple enough to describe. Metabolism, growth, adaptation, reproduction, responsiveness to stimuli, learning and movement, are just some of the terms that have been used to describe it, and there are many more. But the problem of defining life in this way is that it is rather cosmetic. It does not get to the core of what makes life really special. To discover its secret we need to get below the skin of these attributes and find a common denominator. On first sight this might appear to be a daunting prospect since a huge variety of life exists on Earth, which some estimates put at over 30 million species. However, in spite of this great diversity, it has been shown that every form of life, from the most simple to the most complex, shares certain characteristics. Put simply, it can be argued that all life: (1) is composed of one or more individual units called cells, (2) has evolved from a common ancestor, (3) is capable of reproduction, and (4) contains genetic material in the form of a very special chemical called deoxyribonucleic acid (DNA).

In fact, this list can actually be whittled down even further because the first three attributes are a direct consequence of the fourth. In other words, the one fundamental thing that unites all life is DNA. Every living thing that has existed over the last 4 billion years or so has contained at least a small amount of DNA, and without it life would be impossible. The reason is simple: DNA provides life with its two special and almost unbelievable properties – the ability to replicate itself and the ability of being able to take non-living substances and to transform them into living matter. By any stretch of the imagination, these two processes are remarkable, and fundamental prerequisites for life.

AN INTRODUCTION TO DNA

DNA (which will also be discussed in chapter 4) is just a chemical, but it is probably the most complex and astonishing chemical in the whole universe. It consists of two long strands that are wrapped around each other in the shape of a double helix (and is normally to be found coiled-up tightly and packed into chromosomes). Holding the two stands together as they swivel around each other are pairs of simple molecules (like rungs of a ladder) known as bases. DNA contains only four types of base (adenine, guanine, cytosine and thymine) and because these are held together with weak bonds, the two strands making up the DNA are easily able to 'unzip' and separate into two units. The bases also have another important characteristic in that they are very selective in who they form bonds with. In fact, adenine can only bond with thymine, and cytosine with guanine. Consequently, when the two strands of DNA unwind and separate, each of the individual 'exposed'

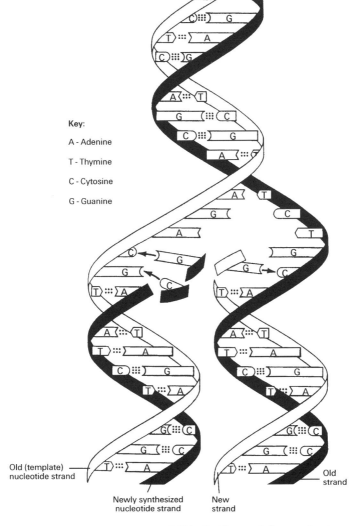

Figure 2.1 Diagram showing how DNA divides and forms the template by which new copies can be made (replication).

bases can only act as a magnet for its own 'special' or complementary partner. If the exposed bases are successful in attracting new partners then the result is the construction of a brand new strand that is identical to the old one. In this way, one molecule of DNA is able to transform itself into two (see figure 2.1).

This would be remarkable by itself, but DNA has another special property – it is able to create life. Encoded in the long DNA molecule are large numbers of genes (actually made-up of long sequences of bases) that are the starting point for making a huge variety of chemicals (or more specifically proteins) which are crucial for life. A single protein by itself is not 'alive' but DNA breathes life into the proteins it produces by using them to construct self contained microscopic worlds called cells, which form the most basic unit of life (e.g., the simplest forms of life are single-cell organisms). Cells (and their DNA) also require energy and various other chemicals from the outside world. Nevertheless, they are able to maintain themselves as independent units, and more importantly provide a protective environment for the all-important and delicate DNA. The only problem is that every time DNA makes a new copy of itself, it also has to go to the trouble of constructing another little cellular world for itself to live in.

DNA has been likened to a blueprint that contains the instructions for a cell to be constructed from scratch. But, of course, advanced organisms like ourselves are multicellular and contain many different types of cell. If we didn't know better, we might expect that each type of specialised cell (neuron, muscle, skin cell, etc.,) would contain its own unique DNA. But nature has not designed things this way. In fact, all of our cells contain exactly the same DNA – the blueprint for constructing the *whole* organism is to be found wrapped up in the chromosomes of practically every cell in the body. A muscle cell is different to a skin cell not because of any differences in their DNA content, but because a different pattern of their genes has been switched on and off. The process in which rudimentary cells turn out to become specialised units is known as differentiation and its importance can be shown by the fact that all life starts out as a single cell with just one set of chromosomes. In the case of humans, this single cell is a fertilised egg that contains a mixture of genes taken from both parents. Given the right set of conditions, the egg will begin to divide into two cells, and then four, and so on. This process in turn will lead to new and different cells with specialised functions and result in the formation of an embryo with different parts, and then finally an individual. But no matter how many cells are produced they will always contain the original compliment of DNA as found in the very first cell.

The fact that DNA insists upon being passed from parent to offspring reveals a fascinating fact about life. As Richard Dawkins points out in his book *The Selfish Gene*, our genes are essentially immortal because they do not change (unless by accident or mutation) as they are passed on through new generations. Put another way, we carry around exactly the same *types* of genes as our ancestors did many thousands of years ago. Indeed, the vast majority of our genes have passed through thousands upon thousands of generations over millions of years without being significantly altered. (Changes or mutations in gene structure are normally harmful, although in some situations they are believed to be the spur that leads to a creation of a new species). Thus, genes have in effect achieved a state of immortality; just as long as they can be passed from cell to cell, or in the case of multicellular organisms, from individual to individual.

THE ORIGINS OF LIFE

The origin of life, and therefore the formation of the first genetic material capable of replicating itself, is believed to have occurred some 4,000 million years ago in the Earth's so-called primordial soup. The vital breakthrough was the creation of unique compounds called nucleic acids (the building blocks of DNA). Individually these are simple chemicals, but they have the important ability of being able to join themselves together to form long chains that make more complex molecules. In fact, nucleic acids are so simple that they only contain three types of chemical: a phosphate, a type of sugar (which in DNA is called deoxyribose) and one of the 4 bases. These chemicals in turn are made up of just phosphorous, carbon, nitrogen, hydrogen and oxygen. In fact, out of the 92 elements that occur in nature, DNA is composed of just five and, with the exception of phosphorous, the other four elements are the most abundant elements in the earth's biosphere. Moreover, since all elements have a tendency to combine together to form new substances, it is perhaps not surprising that the accidental formation of nucleic acids occurred at some point in Earth's history. And once formed, it was probable that they would join up with other nucleic acids to make more complex forms – including, of course, DNA.

As we have already seen, one of the special features of DNA is its ability to replicate and therefore confer immortality upon itself. But there was a problem in the primordial soup. DNA was (and still is) a very delicate molecule. It is easily damaged by such things as heat, radiation and chemical exposure, and consequently it is certain that the first forms of DNA were damaged almost as soon as they were created. In order to survive it had to find a way of protecting itself, and it did this by creating for itself a protective jacket which in turn led to the development of what we now recognise as the cell. But of course, this jacket requires constructing. The material which DNA utilised (i.e., took from its watering surroundings) in its grand design were amino acids which can be combined to make extremely complex three-dimensional molecules called proteins. Therefore, for life to begin there must have been both nucleic acids and amino acids floating around in the primeval soup. And, at some point DNA made the giant step of utilising amino acids to create its own little world.

Until the 1950s this theory had been mere speculation for the simple reason that nobody had been able to demonstrate that even the most basic chemicals of life were able to arise from inanimate material. One problem in attempting to prove that such a development was feasible is the fact that the Earth's atmosphere today (which is rich in oxygen), is very different to that of 4,000 million years ago which was largely derived from volcanic vapours and consisted mainly of nitrogen, methane, water, and hydrogen. Although oxygen is vital for life that exists today, it is not conducive to the formation of nucleic acids. Consequently most scientists believe that it is very unlikely that nucleic acids could arise spontaneously in today's conditions. But the situation some 4,000 million years ago was very different.

In 1953, Stanley Miller and Harold Urey decided to manufacture, under laboratory conditions, the type of atmosphere that probably existed when life first made its appearance. They created a miniature enclosed world in which a warm flask of water simulated the primeval sea, with an atmosphere that consisted of methane and ammonia and circulating water vapour. From time to time an energy source was added to the atmosphere by generating electrical sparks to simulate lightning. After a few days the solution in the flask turned from clear to murky brown. More importantly, after just one week, Miller and Urey analysed the contents and found a number of organic compounds including an amino acid. Variations of this experiment performed by other scientists have produced other amino acids; purine and pyrimidine bases as found in DNA; and several types of sugar. This research strongly suggests that all the building blocks essential for creating life were in place and waiting to be utilised in Earth's early primordial soup.

AGING AND DEATH IN BACTERIA

Although the chemical bits and pieces necessary for life were probably floating around some 4 billion years ago, it took another 500 million years for the first forms of life to be created. As far as can be established from fossil evidence, the earliest types of life were bacteria which appeared some 3,500 million years ago. Bacteria are essentially microscopic single cells that contain no nucleus (see below) and consist of a small amount of genetic material existing as a loop in the cell's watery interior or cytoplasm. The technical name for these types of cells is Prokaryotic (*pro* = before, *karyo* = nucleus) and even today these simple organisms are the most abundant form of life on Earth. Furthermore, prokaryotic cells and bacteria were the only kind of life on Earth for the next 2,000 million years, during which time they flourished and evolved into millions of different forms.

It is at this point in evolutionary time that we may have expected natural aging (and death) to have made their first appearance. But it seems that this is not the case. Biologists have found that bacteria do not appear to age or grow old and they only seem to die through starvation, accidents or genetic damage. Therefore, bacteria appear to have the potential to be immortal, and the same is true of many other one-celled organisms such as algae, amoebas and certain types of protozoa. Aging and death, it would seem, are not inevitable consequences of life. However, there is a catch: the main reason why bacteria do not grow old is because they rejuvenate themselves by replicating. Bacteria normally reproduce asexually (i.e., by themselves) via a process of binary fission in which they split down the middle and divide into two. The 'old' cell in effect disappears and becomes incorporated (or rather transformed) into two new ones. Almost as if by the sleight of a magician's hand, the old is turned into new, and aging as we understand it does not occur. In an evolutionary sense this is immortality, although from our own human perspective it looks very much like cheating.

MULTICELLULAR ORGANISMS

About one thousand million years ago a new type of cell emerged from the primordial soup known as the Eukaryotic cell (*eu* = true, *karyo* = nucleus) which was a much more complex and advanced world that the DNA had created for itself. Firstly, eukaryotic cells contained a centralised control centre, or nucleus, that contained incredibly huge amounts of DNA. Secondly, the interior (cytoplasm) of the cell contained a number of additional structures with specialised functions. Two of these structures were the mitochondria (which provided the cell with its own energy source), and ribosomes (which enabled the cell to make its own complex proteins using information coded in the genes of the DNA). One consequence of this new development was that the eukaryotic cell began to include within its DNA for the very first time the code for constructing many different types of cell. The result was the development of multicellular organisms, and it was probably at this point in evolutionary history that aging and death finally became a characteristic of living organisms. To understand why, we need to look more closely at the structure of multicellular life.

The human being is a good example of a multicellular organism. It has been estimated that the human body contains over 100 million million cells, which can further be divided into many different types (muscle, skin, nerve cell, etc.,) each with its own specific structure and function. But where do these cells come from? In one sense (as mentioned earlier) they all ultimately derive from a single egg (ovum) that has been fertilised by a sperm. The ovum and sperm, of course, provide the bridge by which genetic information is passed on into a new individual, and for this reason they are known as germ cells (deriving from the word 'germinate'). This is a particularly apt description of the fertilised egg which contains all the genetic information required to create a new individual. In order to do this, however, the fertilised egg must give rise to many other types of cell that lead to the formation of the body of the new embryo or individual. Obviously, the vast majority of these cells are not designed to pass on their genetic inheritance, but are responsible instead for creating the main structure of the organism. These types of cell are consequently called *soma cells* (meaning body) to make them distinct from *germ cells* (sperm and egg involved in genetic transmission).

It was pointed out by the German biologist August Weismann, back in 1891, that germ and soma cells are also different in terms of their mortality. Germ cells are designed to pass on genetic information; but soma cells are a genetic dead end. For example, a muscle cell has no mechanism to pass on its genetic information to a new generation despite the fact that it contains the same genes as the original egg from which it is derived. Thus, it does not matter one jot in an evolutionary sense whether the muscle cell (or indeed any other soma cell) is mortal or immortal. In other words, once the body cell has done its job it can be allowed to die. The same is not true, however, for a germ cell, which is required to pass on genetic information and has to be potentially immortal so that its contents can continue in a new individual.

Clearly, genes have to remain unchanged from generation to generation otherwise the species that they create would soon become extinct (which would be an absurd waste of time considering the millions of years it took for them to be formed in the first place) Therefore, with the development of multicellular creatures, a situation arose where soma cells could be allowed to age and the organism die – providing of course the organism managed to successfully reproduce and pass its genes on to its offspring. In this way genes remain essentially immortal, whereas the individual dies. In short, the cost of specialisation into germ line and soma cell is death.

This scenario raises the question of whether soma cells were once immortal like their germ cells counterparts. A number of theorists have assumed that initially they were; but since immortality was not essential, soma cells have become subject to a different set of evolutionary pressures where new trade-offs and benefits made them mortal and finite (a variation of this idea is discussed later as the disposable soma theory). As yet, there is no definite proof to support this idea, but the suspicion exists that normal or healthy soma cells once had, and maybe still have, the potential to be immortal.

Thus, when placed in an evolutionary perspective, aging and death can be linked to the development of multicellular organisms and, in particular, to the different mortalities of germ and soma cells. Another important evolutionary development was the occurrence of sexual reproduction. Biologists have long argued over the reasons why sex should have evolved, although there is no doubt that one of its great benefits is the great variability of gene combinations it throws up compared to asexual reproduction. This constant shuffling of genes, and the large number of different individuals it creates, provides variation and great evolutionary advantage to the species as a whole. Some researchers have claimed that aging has evolved in conjunction with sexual processes. Strictly speaking, this is not absolutely true as some single-celled organisms such as algae are able to reproduce sexually without apparently showing signs of aging. Nevertheless, the development of sexual reproduction did increase significantly the genetic and multicellular complexity of life, and was undoubtedly an important factor in the evolution of aging.

To recap: for life to continue and replicate itself, its genes have to be immortal. Or at least immortal in the sense that copies of the genes must be able to be passed on from parent to offspring, for all intents and purposes, for ever. From the gene's own perspective, as long as identical copies of themselves are passed into a new body, then the soma cells that are left behind are disposable. And it is probably for this reason that aging and natural death first emerged in living organisms. In short, it doesn't matter if the individual dies, just as long as the genes are passed on before this event occurs. Another way of looking at the same situation is to assume that death occurs because soma cells cannot maintain (or renew) themselves for ever. In other words, there is a limit to how long they can survive. Germ cells are different. An almost endless stream of genetically identical germ cells (sperm and ova) are continually produced by both sexes, and it only takes one of these to become involved in fertilisation for its genetic material to renew itself and effectively escape from the sentence of death.

Species Longevity

Although aging may be the inevitable price that has to be paid for the development of multicellular life, this explanation also leaves many important questions unanswered. Not least is the fact that soma cells vary greatly in their life span which is reflected in the tremendous variation in aging that exists between living organisms. For example, in mammals, the maximum life span of species varies from about 1 year in the smoky shrew, to about 3 years in the mouse, to 40 years in the horse, 70 years in the elephant, and approximately 120 years in the human. If considering the animal kingdom as a whole then the range becomes even greater. The mayfly has a life span of less than 24 hours, whereas the Galapagos tortoise may live up to 200 years (one adult Galapagos tortoise of unknown age, captured in 1766, lived for 152 years in captivity before dying after a serious fall in 1918). In fact, there may be animals with even greater life spans. Amongst the most fascinating are the large slow growing fishes, such as sturgeon, that continue growing for as long as they live. A sturgeon weighing 1,500 pounds has been officially recorded that has been estimated to have been at least 200 years old when caught; and a type of deep sea fish known as the behemoth has tipped the scales at over 3,000 pounds which suggests an even greater longevity.

From one day to over two hundred years! It is hard to account for these age differences simply in terms of soma cells wearing out. Presumably, some soma cells age faster than others, and consequently there must be other factors that determine how long the organisms will survive. Indeed, each and every species seems to have its own individual life span, and it is hard to escape the conclusion that to some extent it must be genetically determined. But how can the genes affect the life span of soma cells? Perhaps the most fundamental way of tackling this question is to explain how and why aging has evolved as a feature of multicellular life. The great variability in aging strongly indicates that aging must have evolved for very good reason. To understand why, we need to examine the concept of evolution in more detail.

CHARLES DARWIN AND EVOLUTION

The concept of evolution starts with Charles Darwin's book *On the Origin of Species by Means of Natural Selection*, published in 1859 and widely recognised as one of the most important books in the history of civilisation. Darwin was just 22 years old when he sailed from England on the H.M.S. Beagle in 1831. The primary mission of the five year voyage was to chart poorly known stretches of the South American coastline, although Darwin's task as the ship's naturalist was to collect new specimens of fauna and flora as he went on its journey. During the voyage Darwin was struck by the great diversity of life he came across, particularly that which he found on the relatively small and isolated islands of the Galapagos. Despite the diversity, Darwin also noted that many of the species also shared certain similarities. For example, the islands supported many different kinds of finch that shared common features, yet each was uniquely adapted to their own particular habitat. It was almost as if all these birds had arisen from a common ancestor, but had become slightly modified to allow them to adapt more effectively into their own special ecological niche.

This may sound like a perfectly reasonable theory today, but 150 years ago it was in direct contradiction to the traditional teaching of the church, which held that all species, including man, had been made by the divine hand of God, and had remained unaltered since the time of creation. The implications of Darwin's observations were both revolutionary and blasphemous as they contradicted long-held and cherished religious views. Indeed, he knew that his theory would be highly controversial, and partly because of this, Darwin spent more than 20 years working on his ideas before daring to publish them. The impact of the work was nevertheless highly sensational, and almost at a stroke shattered mankind's notions of himself and his place in the universe. Its impact on science was just as great since it helped to unlock the mysteries of life and launch modern biology. Arguably, no other book has ever had such an impact on how we understand ourselves and the world around us.

The suggestion that all living things change with time, which is the fundamental notion of evolution, did not originate with Darwin, but he was certainly the first to provide a plausible and convincing mechanism by which it could occur. The theory he proposed was called Natural Selection and was based upon two simple concepts: competition and variation. The first concept derived from Darwin's observation that all living organisms appear to produce more offspring than are needed to replace their parents, even allowing for the fact that some offspring are killed and lost through accidents. The consequence of this effect is that too many individuals are produced for the resources that exist (even slow breeders such as elephants produce more offspring than their environment can adequately support). Yet it is clear that animal populations tend to remain relatively stable and do not expand beyond certain limits. The consequence of increased numbers of offspring must therefore be that all creatures are thrown into competition with each other.

The second component of natural selection is that all individuals of a species show great variation in terms of their biological characteristics. For example, as humans we may all look similar, but every one of us is different in terms of physique, strength, intelligence etc., and the same basic premise holds true for other animals. The consequence of this variability is that some individuals will be much better suited to their competitive environment than others. The individuals that are best suited will in turn, be the ones most likely to reproduce and pass their genes on into their offspring. In this way the 'fittest' individuals (that is those best adapted to their environment) will be the ones whose genes are passed on to continue the species. This process is natural selection. However, Darwin's most important insight into natural selection was his realisation that over the course of many generations the selection process would undoubtedly cause great changes in form to develop. For example, because ancestral giraffes had a liking for feeding in tall trees, natural selection would favour the development of long necks. Following this argument to its logical conclusion, it was but a short step to predict that the ultimate development of natural selection was eventually the origin of a new species.

Although Darwin described the process of evolution in detail, he did not explain how inheritance worked. Genes had not been discovered in the 19th Century, and the lack of a cogent explanation of how inheritance operated was the main obstacle to his theory. Ironically, such a theory had been formulated in 1865 by a young Augustine monk living in Bohemia called George Mendel, but the work had not been fully appreciated at the time and was forgotten until it was re-discovered in 1900. Mendel's theory provided the precise rules of genetic inheritance that Darwin's theory needed to explain how natural selection worked.

Sadly, Darwin died in 1882 and never knew of Mendel's research or the true legacy of his ideas to biology.

THE WORK OF AUGUST WEISMANN

Returning to the central question posed at the beginning of the chapter of why each individual species has its own characteristic life span, one answer is that it has evolved that way. The first evolutionary theory of this sort was developed by August Weismann. As we have already seen, mortality appears to be a property shared by all multicellular organisms, and Weismann was the first to speculate that aging and death went hand in hand with the evolutionary development of somatic cells. In short, he saw death occurring because soma cells were disposable, and because they had no means of renewing themselves. But Weismann was also unhappy with this idea as the full explanation because it did not explain why different species had such diverse life spans. Looked at from another perspective, however, it also seemed to Weismann that without death, the world would quickly become over-populated, leaving no resources (or space) for a species to evolve. In other words, life without death would be evolutionary disaster. Thus, he reasoned that death must have evolved as a result of natural selection in order to eliminate the old and worn out members of the population. Weismann did not explain how this process occurred in evolutionary terms, but the idea made intuitive sense, and presumably somewhere in this theory was the reason why different animals had such widely different life spans.

Unfortunately, the problem with Weismann's theory is that there doesn't appear to be any possible way of explaining how aging and death could have evolved as a result of natural selection. In fact, it must be the case that natural selection (i.e., the process resulting in the survival of the fittest individuals from the group or population) can only act on individuals, and there is no way it can operate indiscriminately on the species as a whole. Moreover, it is also clear from the individual's own viewpoint that there is no benefit at all to be gained from death. In fact, natural selection must work to promote fitness (which is the opposite of death), and thus it should always work to increase the chances of life rather than decrease them. A second problem for Weismann's theory is the fact that very few wild animals ever reach the end of their natural life span since they die or are killed beforehand. Therefore, there seems to be no way in which evolution can select for longevity in individuals. Weismann had started the ball rolling in an attempt to find an evolutionary theory for aging, but it was clear that if evolution was indeed involved in determining life span, then a very different type of explanation would have to be found.

SIR PETER MEDAWAR

A second theory of how aging may have evolved was proposed by the Nobel Prize winning biologist Sir Peter Medawar during the 1940s. In accordance with Darwinian theory he supported the idea that natural selection worked by favouring 'fit' genes and weeding out the weak. However, Medawar also pointed out that natural selection cannot operate once an animal has reproduced. This is because once reproduction has occurred, the animal has in effect proven itself 'fit', and consequently its 'successful' genes are passed-on to face natural selection in a new organism. Thus, natural selection can only act on genes that are expressed before reproduction. Since most genes are operational during early growth and development then clearly the vast majority of genes will indeed fall under the influence of natural selection. This may also have other evolutionary benefits since any damaged genes that spontaneously arise early in life will invariably result in an individual that fares badly in the reproductive stakes, and whose genes are not likely to contribute further to the species gene pool. However, Medawar also argued that there may be a few genes that escape natural selection because they become switched-on late in life and after reproduction has occurred. If these genes turn out to be harmful, they are likely to remain so because they will not have been exposed to the unforgiving forces of natural selection. One probable consequence of this situation is that there will be a drift of harmful genetic effects towards the end of life, resulting in an increased probability of death.

Some support for Medawar's theory comes from the genetic disease Huntington's Chorea first described by George Huntington in 1872. This disease is characterised by complex and involuntary jerky movements from which the disease gets its name ('chorea' derives from the Latin for 'dance'). Along with the disruption in movement is a progressive dementia often resulting in a terminal state of physical and mental helplessness. There is no cure for the disease and generally death occurs within 10 years of onset. However, what makes the disease particularly tragic is the fact that the disorder doesn't normally start to manifest itself until about 35 to 40 years of age. Consequently by the time symptoms start to appear, most carriers will have had children (who in turn run the terrible genetic risk of having a 50:50 chance of inheriting the disease).

Because the gene does not express itself until after reproduction (for reasons that are not known) natural selection has not been able to eliminate it, and therefore it continues to exist in human populations. Recent research has shown that the disease is due to a mutated gene located on the short arm of chromosome 4, and this discovery has enabled the disorder to be diagnosed with a high degree of accuracy using DNA testing. Whether people want to know whether they are carrying a particularly unpleasant incurable disease, however, is another matter.

The biggest problem with Medawar's theory is that one would expect a large build up of genetic diseases with aging although this does not appear to occur. In fact, the emergence of new genetic disease in old age is relatively rare, and genetic disease certainly does not appear to be a major cause of death. Of course, it might be that the genetic effect is more subtle, but there is little evidence to support this idea either. Criticisms aside, Medawar's theory does lead to some very interesting theoretical predictions. For example, if we wanted to increase the human life span, one possibility might be to ban reproduction before a certain age, say 40 years, and then to progressively raise this threshold with time. This strategy would presumably 'expose' the bad genes before reproduction occurred thereby eliminating them through natural selection. A second strategy might be to 'fool' the genes into thinking that the body they occupy is younger than it really is. The mechanism by which some genes become switched-on and others turned-off is not fully understood, although it may well involve certain chemical messages, some of which are hormonal in origin. Finding ways of maintaining the body in a youthful hormonal state, therefore, might perhaps be expected to help delay the expression of any late-acting genes.

PLEIOTROPIC (DUAL FUNCTION) GENES

An important modification of Medawar's late genetic build-up theory was published by George Williams in 1957. He pointed out that many genes are actually *pleiotropic*, which means that they have more than one effect. An example of this type of gene is the one that causes muscular dystrophy. This condition is nearly always found in males and is characterised by muscle wasting and weakness which first appears in early childhood, and progresses rapidly so that the victim is typically confined to a wheelchair by early adolescence. Sadly, survival rarely extends beyond the second decade of life. Muscular dystrophy is known to be caused by a mutated gene that is located on the sex-linked X chromosome, but this lethal gene does much more than just cause muscular damage. It also leads to cataracts, diabetes, malfunctioning of the testes, irregular heart beat and mental retardation. Thus, the muscular dystrophy gene is pleiotropic because it manifests itself in many different ways.

It is probably the case that the vast majority of genes, if not all, have some type of pleiotrophic effect, although some are more pleiotropic than others. A variation on this basic idea was used to explain aging by George Williams. He argued that although most genes were likely to have evolved in order to produce beneficial effects early on in life, it was feasible that the same genes could have adverse consequences later. This effect he called *antagonistic pleiotropy*. Williams illustrated the idea by using an hypothetical example of how our bodies might be programed to use calcium. During childhood and adolescence, calcium metabolism might be particularly efficient in bones allowing them to grow quickly or to heal following damage. However, the same gene may also be partly responsible for the slow but steady calcium deposition in the arteries that accompanies the development of heart disease. Such a gene might be chosen by natural selection because most individuals will benefit from its advantages in youth, while few will live long enough to experience its disadvantages. Even if this gene led to certain death by 50 or 60 years, it is still likely that it would be favoured by natural selection as long as it produced some early benefit.

Although there is no clear evidence that calcium deposition is a real life example of antagonistic pleiotropy (although it is entirely feasible), there are other examples that have been used to support the theory. For example, some children are genetically predisposed to be overweight, because they are very efficient at turning calories into fat tissue. This innate tendency could actually be an advantage in many countries where there is a poor or fluctuating food supply. But of course, this same trait is likely to predispose individuals to diabetes and heart disease later in life in industrialised countries, where there are no food short-ages. This example has sometimes been cited as an example of antagonis-tic pleiotropy since efficiency in storing fat may be a potential advantage in youth, but is likely to become detrimental as the person gets older.

One of the main predictions of the antagonistic pleiotropy theory is that genes (or rather their biological effects) with early benefits will have significant costs attached to them later on in life. In evolutionary language this means that genes favouring fitness in youth will be at the expense of fitness later on. George Williams has summed this up provocatively by saying:

> *It follows that an individual can not be exceptionally gifted with both youthful vigour and long life. I would predict that*

> *no human being who is over a hundred years old*
> *was unusually vigorous as a young adult.*

Although Williams does not explicitly define vigour it is nevertheless implicit that he is referring to reproductive fitness, and the implication is that the early onset (or capability) of sexual activity is not conducive to longevity. As a potential test of this hypothesis it would obviously be intriguing to know whether centenarians, for example, were individuals who indeed started their sexual activity at a later point in life than those who are shorter-lived. Unfortunately obtaining this type of information is fraught with methodological difficulties and evidence relating to this most interesting issue appears to be non-existent.

Research relating to this issue has, however, been performed in animal studies and has produced good evidence showing that genes with early benefits do appear to contribute later to reduced fitness and senescence. One approach has been to take a species with a short life span (so that the study does not take too long) and to selectively breed individuals that start to reproduce early in their life cycle. The offspring of these animals are in turn bred, and the first reproducers among them selected again for breeding, and so on. Using this strategy it is possible to breed a group of animals who consistently show very early reproduction. This type of study has been performed by David Mertz working at the University of Illinois using flour beetles (*Tribollium*) and with striking results. For example, after just 12 generations, the average number of eggs produced during the first month of life increased from 417 to 460, while length of life (in males) decreased from 271 to 231 days. This would seem to lend excellent support to Williams' theory mentioned above. Moreover, studies using fruit flies have examined the consequences of selecting animals to reproduce later in their life cycle. In accordance with the antagonistic pleiotropy theory, fruit flies that produced more offspring later in life also tended to live longer (although in total they also produce fewer offspring than their younger counterparts).

THE DISPOSABLE SOMA THEORY

Medawar's theory of late-acting genes, and the more likely concept of antagonistic pleiotropy, both propose that the life span of every species is the result of natural selection acting to maximise fitness in the early part of life. But, one way or another, this inevitably has long term disadvantages leading to decline once the reproductive period is over.

In other words, aging is seen as a by-product of selection for benefits that occur early in life. Although both these theories provide an evolutionary explanation of why aging occurs, they still do not really get to grips with the important central issue of why different species have their own characteristic life spans. Inevitably, there have to be other factors involved.

One important theory has been developed by Thomas Kirkwood. He points out that any given organism can be likened to a 'black box' whose most important objective is to take energy from the environment and to use it to produce offspring. However, to achieve this aim, the organism must also use some of its energy to cover other important requirements. For example, the organism must allocate some of its energy to growth and some to protecting itself. In addition, the organism must also invest considerable energy into making sure that its genetic material (or the germ line) remains functional and intact. Furthermore, a certain amount of energy also needs to be invested into producing and preserving all of the body's soma cells. Despite these various demands, energy is a limited and finite resource and there is only so much that can be used. Consequently, energy invested into producing one benefit will be at the expense of something else.

According to Kirkwood, one of the most important factors in this energy equation are the soma cells. As mentioned earlier, soma cells, unlike germ cells, are mortal because they are essentially genetic dead ends. But this is a rather simplistic explanation. In fact, one reason why germ cells are immortal is that the organism has to invest considerable energy into maintaining the integrity of its genetic material. Indeed, it is known that germ cells have highly specific enzyme systems and proof-reading devices that enable the DNA to replicate itself with high fidelity, and if this did not occur then the survival of the species would be in jeopardy. Although soma cells also require repair and maintenance, particularly in long lived species, the same degree of investment as in germ cells would be highly wasteful and confer no real advantage. Thus, the difference in mortality between germ and soma cells can be attributed to the different energy investment the organism has to make in maintaining these cells. In other words, the organism *has little option* but to expend considerable energy to maintain the immortality of the germ line, but the same degree of investment is not necessary for the soma cells.

In an evolutionary sense, therefore, soma cells are disposable. This does not mean, however, that they are unimportant. On the contrary, if any given species was to evolve a longer life span, then evolution must have achieved this aim by somehow acting to increase the longevity of the soma (body) cells. In other words, considerable energy must have been directed into making sure soma cells could survive for long periods. In fact, as we shall see later, soma cells have indeed evolved a number of important defence systems against aging. One of the most important (and this will be discussed at several places throughout the book), is the defence against oxygen and its chemical reactions. We may need oxygen for life, but the break down products of oxygen (known as free radicals) are highly damaging to biological tissues and consequently our cells have evolved a number of enzymes (including superoxide dismutase, catalase and glutathione peroxidase) that help to reduce oxygen's harmful effects (see also pages 67–68 and 113–118). Soma cells have also found other ways of extending their hold over life. For example, many cells of the body can be replaced and renewed many times over once they become old or inefficient, and it has been shown that this replicative capacity is far greater in long lived species (chapter 6). Moreover, longer-lived species also have more efficient DNA repair mechanisms that help to correct any genetic damage that takes place (chapter 4). In short, soma cells have evolved an arsenal of protective weapons against aging and these provide an important key to understanding how different aging characteristics have evolved between different species.

But if soma cells have evolved in this way, then what has been lost in the trade-off? According to Kirkwood, increased soma cell investment will mean that there is less for reproduction. Obviously, the organism must strike a fine balance between the two requirements and Kirkwood argues that the extent of this balance will depend on the species and its own particular ecological niche. For example a species subject to high accidental mortality will do better not to invest heavily in soma cells, but should depend instead on more rapid and prolific reproduction. If this strategy is adopted then the animal should produce lots of offspring but be short lived. Alternatively a species with low levels of accidental mortality (like ourselves) may profit by the reverse strategy – that is, to increase longevity at the expense of reproduction. This theory does indeed appear to have considerable support. For example, mice which start to breed at about 6 weeks can produce several litters in a year and their life span is about 3 years; domestic cats start breeding at about 1 year, produce 2 or more litters annually and have a life span of 15–20 years; and horses typically have one foal a year and live 20–30 years, and so on.

Kirkwood's theory not only provides an evolutionary explanation for aging, but it also helps explain why aging occurs on an individual basis. In short, according to Kirkwood: "aging should be the result of the accumulation of unrepaired somatic damage". In other words, once the biological processes that help to maintain somatic cells start to become inefficient and wear out, then aging and death will ensue. This theory is also generally regarded as a variation of the antagonistic pleiotrophic theory since any genetic involvement in enhancing the maintenance of somatic cells early in life may (in theory at least) be expected to have drawbacks later on.

GEORGE SACHER AND LONGEVITY ASSURANCE GENES

The basic thrust of all the theories discussed so far is that the decline of aging is bought about by the result of adverse genetic or pleiotropic effects that tend to accumulate late in life. In addition, although it is not immediately obvious, these theories also rest on the assumption that; in the beginning, both germ and soma cells were immortal, but because of the evolutionary benefits, soma cells were sacrificed to become mortal. Viewed this way, aging and death can be seen to be something that has emerged relatively recently in evolutionary history, from a 'perfect design', and occurs because, in some way or another, the processes leading to its development have been advantageous to the species. However, this position is not accepted by everyone. For example, the late George Sacher believed that soma cells were, and always have been, finite in their life span, and this led him to propose that living organisms did not give up on immortality, but rather they had to fight every inch of the way to gain it. Viewed in this context, length of life is not due to the occurrence of late acting senescent genes, but rather because evolution has favoured the development of genes that favour longevity.

This idea of aging can perhaps be illustrated using an automobile analogy. Because motor manufacturers design their cars to last a certain period of time without serious malfunction, they typically provide new vehicles with a limited warranty against breakdown. However, once the warranty period is over the guarantee ends, and although the car may run perfectly for several more years, systems within the car will gradually decline which makes their failure at some later point inevitable. The same may also said of living organisms. Our own warranty extends to early adulthood, and once this period is over then our body systems begin to decline. Sacher would probably argue that our warranty period is governed by longevity assurance genes that ensure that the organism is protected against eventual biological breakdown for as long as possible. But, they can't last out for ever and their eventual weakening of function consequently results in aging.

Indeed, Sacher sees no need to invoke the concept of antagonistic pleiotropy in this explanation. In his view, aging is simply the result of decline in the body's biological systems and need not have a specific genetic basis.

As already mentioned, there is a huge range (over a hundred-fold difference) between mammals in terms of their maximum life spans; and rather than trying to explain these differences in terms of late-acting senescent genes, Sacher set out to see if he could discover other evolutionary factors that appeared to be linked to longevity. He attempted to do this by trying to establish whether differences in mammalian longevity were correlated with any particular anatomical or physiological characteristic. Sacher started by examining the life spans of 85 species ranging is size from shrews to elephants. One of the first trends to emerge from this data was that the larger animals tended to be the ones that were longer-lived although this relationship was far from

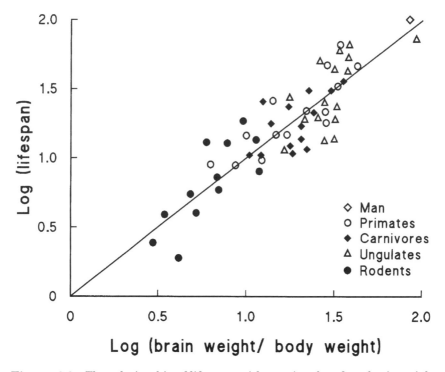

Figure 2.2 The relationship of life span with a ratio taken from brain weight over body weight for various mammalian groups (data redrawn – not to scale – from Lamb, M.J. (1977) *Biology of Aging*. Blackie and Son. Glasgow.)

perfect. For example, man is the longest-lived mammal and is considerably smaller than the elephant or rhinoceros. Sacher then decided to look at brain size and found that animals with larger brains also lived longer. But again, humans with their relatively light brains were an exception and did not fit the general pattern. Finally, Sacher decided to adopt a new measure by using a ratio of brain weight to adult body weight. Using this score, Sacher found that the larger the brain in relation to body weight, the longer the species life span. Furthermore, all of the species he examined (including man) fitted this pattern with a high degree of accuracy (see figure 2.2).

In fact, these findings correspond well with observations that have long puzzled people who study records of domestic animals. Small sized dogs live longer than larger dogs, and they have a larger brain in proportion to total body weight. Ponies have a relatively larger brain in comparison to body size than horses, and sometimes they may live to be 40 years, whereas 34 is extraordinary for a horse. Even the longer average life span of women compared to men may possibly be due to the differences in the relative growth of brain and body. Between birth and adulthood, a boy's body weight increases 4.6 times faster than his brain; whereas a girl's body grows only 4.2 times as fast as her brain, and the result is that she has a bigger brain in relation to her body. In short, it appears that a large brain placed into a small body is an excellent indicator of longevity.

What has this got to do with evolution? Well, it appears that evolution in mammals has, in the main, progressed towards evolving species with large bodies and big brains. And in turn, this has resulted in animals with greatly increased life spans. These findings according to Sacher do not support the idea of late-acting senescent genes, but rather in his view, it suggests that life span is closely linked to the development of certain structural characteristics. Interestingly, Sacher has also estimated from fossil records that over the last 200,000 years the size of the human brain has increased by 12%, and that there has been an average 40 year increase in life span. In evolutionary terms this is a very short period of time, and he has further estimated that over this period probably less than 250 genes would have been altered to produce these changes. The majority of these genes would probably have been involved in causing structural change resulting in increased body and brain size etc. However, a few may also have been specifically involved in increasing life span (perhaps by helping to improve DNA repair in somatic cells for example) and therefore qualify as longevity assurance genes.

The two main evolutionary theories of aging covered in this chapter (late acting senescent genes versus longevity assurance genes) both have their adherents, although it is probably fair to say that the senescent genes explanation is the more strongly supported. The two theories, however, stand in such contrast to each other that Sacher has argued that at some point in the future, researchers will have to decide between them (although others seem to strongly disagree on this point*). The debate may not be entirely academic since each theory also raises important implications about the potential future benefits of gerontological research. The concept of antagonistic pleiotropy, as we have seen, leads to the conclusion that enhanced early reproductive fitness will have costs attached to it later on in life. However, as Sacher points out, this theory is extremely pessimistic since it implies that youthfulness has its price, and it is difficult to see how there can be any improvement in controlling aging and disease. However, the concept of longevity assurance genes implies that there are no costs to youthful vigour. Indeed, to the contrary, there are clear advantages of increasing youthfulness; and it also opens up the possibility that, one day in the future, life enhancement may be possible through the addition of new or more efficient longevity assurance genes – perhaps occurring through advances in genetic engineering (see also chapter 12).

SUMMARY

- The one fundamental thing that unites all life is DNA.
- DNA consists of two long strands wrapped around each other in the shape of a double helix, that is able to 'unzip' with each strand acting as a template for the formation (or replication) of a new DNA molecule.
- Our DNA (which is packaged in 23 pairs of chromosomes) contains around 100,000 genes that are essentially coded instructions by which to make proteins (chemicals essential for life).
- The simplest unit of life is the cell.
- Aging and natural death probably first arose with the development of multicellular life some 1,000 million years ago.
- Life span must have evolved like any other biological characteristic. The important question is how? Natural selection (as suggested by August Weismann) does not provide a good explanation.

---

*See Rose, M.R. (1991) *The Evolutionary Biology of Aging*, pages 169–170.

- Sir Peter Medawar has proposed that aging is due to the effects of late-acting genes that have escaped natural selection.
- George Williams believed that aging was due to antagonistic pleiotropy; that is, genes with beneficial effects early in life producing harmful effects in the longer term.
- Thomas Kirkwood has argued that animals have not evolved systems to provide soma cell immortality because they have invested their energy instead into ensuring reproductive success (which requires early youthful vigour).
- George Sacher has proposed that life span is due to the evolution of longevity assurance genes.

GENES AND AGING

*The life history of the individual, and his allotment of years,
is the resultant of inborn and environmental influences
inextricably interwoven*

–Louis Dublin

*Possibly in 30 years we will have in hand the major genes that
determine longevity, and will be in a position to double, triple, even
quadruple our maximum life span of 120 years*

–Michal Jazwinski

One of the few things we can say with absolute certainty when it comes to understanding aging is that, one way or another, it has been shaped by the forces of evolution. In addition, because evolution must ultimately exert its effects by altering the genetic design of the organism, it must follow that aging is influenced by genetic factors. Unfortunately, this is where the agreement ends. Nobody knows for sure, despite the many theories, *how* genes affect aging or the *true extent* of their influence. This is much the same as saying that nobody really knows exactly what causes aging other than that genes are somehow involved. But involved they surely are, and most researchers believe that understanding their involvement will in time unlock many of the mysteries of aging. The role of genes in aging can be viewed in two quite distinct ways. Firstly, genes are units of heredity that pass from parent to offspring. This means that if longevity has a genetic basis then, to some extent, it should be inherited from one's parents or ancestors. Secondly, and perhaps more important, is the fact that genes must exert their positive (or negative) effects on longevity by somehow altering the biological make-up of the organism. Both these issues are addressed in the present chapter. However, if we are to fully understand the role of genes in aging there are many other bits of the overall picture we also have to fit in. For example, not only do we have to identify the genes involved; but also explain how they change over the life span, decipher their control over intracellular processes and show how this relates to aging. Taken together this is a staggeringly difficult problem and is probably the most single important area in aging research today. Because of its size and complexity this topic forms the basis of this and, to some extent, the next two chapters of the book.

The fundamental notion of human inheritance is simple; we inherit half of our genes from our father and half from our mother, they get mixed together (in a process called recombination), and the eventual outcome is a new individual who is genetically unique from anyone who has ever existed (with the exception of identical twins). There is little doubt that we inherit many characteristics from our parents and we might expect the same to be true for longevity. Indeed, it has often been said that one of the best ways to guarantee a long life is to have long-lived parents. This idea was summed up over a century ago (before genes were discovered) by the American poet Oliver Wendell Holmes who wrote:

> *The first thing to be done is some years before birth ... to advertise for a couple of parents both belonging to long lived parents.*

This sounds like good advice and few of us would disagree with its logic. But is it really true? Is the length of our life determined by the genes that we inherit from our parents? Of course, in absolute terms, things can never be quite so simple. Our life span is shaped by many other important factors including lifestyle, disease, and accidental happenings, and all of these things will mask any underlying genetic effect. Nevertheless, all things being equal, it should be possible to demonstrate whether there is a genetic basis to aging by looking at the effects of inheritance on large groups of subjects. One way this can be done is through genealogical research in which family trees are constructed and mapped out across several generations. This strategy often involves starting with one (preferably long lived) ancestor, and then tracing all offspring and lineage with exact recording of birth and deaths. Using this approach it is possible to see whether the descendants of long-lived parents do indeed live to greater ages than would be normally expected.

One of the first major investigations to examine the influence of inheritance on life span was undertaken by M. Beeton and K. Pearson in 1899. They examined several collections of old English genealogical records involving thousands of family members, some dating back to the 17th Century. They were particularly interested in measuring the relationship between the length of life of parents and their children, as well as comparing pairs of children from the same parents. To do this they examined the records of over 2,000 pairs of fathers and their adult sons, as well as over 1,000 pairs of brothers. The results showed that, statistically, long-lived parents *tended* to have long-lived children compared to short-lived parents, but the relationship was very weak. Perhaps more striking was the finding that long-lived parents nearly always lived longer than their offspring. Overall, therefore, having long-lived parents held a small advantage, but it was only relative to those less fortunate with short-lived parents.

Another early genealogical study was undertaken in the 1920s by Alexander Graham Bell (the inventor of the telephone), who studied 2,200 male and 1800 female descendants of a person called William Hyde of Connecticut who died in 1681. Bell grouped the age at which members of this genealogy died (using age bands 20–39, 40–59, 60–79, and above 80) and again like Beeton and Pearson found a small but significant relationship between the life spans of parents and their children. For example, when both parents lived to be 80, 20.6% of their offspring reached the same age band; this figure dropped to 9.8% when only one of the parents reached the age of 80; and when neither parent reached 80 years, only 5.3% of the offspring reached this age. Bell further showed that the children of octogenarian parents, on average, enjoyed a life span of 52.7 years which was some 20 years longer than that for the offspring of lesser lived parents. However, again, these results showed that children from long-lived parents often failed to match their parent's ages. Bell also discovered that females tended to live longer than males, although paradoxically it appeared that it was the father's influence, rather than the mother's, that was more important in determining the longevity of this gender difference. However, this relationship has not been supported by more recent studies.

Another important genealogical study was undertaken in 1934 by the famous gerontologist Raymond Pearl who collected information from a large sample of people including some 2,319 persons who had lived past the age of 90 years. Pearl used a measure called "total immediate ancestral longevity" (TIAL) for each person which was calculated by adding up the combined sum of ages at death of their two parents and four grandparents. The results of Pearl's investigation showed that the TIAL was significantly higher for those individuals aged 90 than for the lesser aged groups. For example, 45.8% of those who had attained 90 years had two long-lived ancestors (who had both reached the age of 70), whereas only 11.9% of the control group had ancestors that were similarly long-lived.

Interestingly, Pearl's records from this study were discovered at John Hopkins University in 1960, and over 9,000 offspring of the original group that had lived into their nineties were followed up by a team led by Margaret Abbott. The results, published in 1974, showed that longer-lived parents tended to give rise to longer-lived children although the effect was not particularly marked. For example, the children of parents who had both managed to live to 80 years lived only 6 years longer than children whose parents had both died before 60 years. Abbott and her colleagues also looked at the effects of having one long-lived parent (that had lived beyond 90) along with a shorter-lived parent. Although the results were not conclusive, they indicated that the shorter-lived parent had the greater influence on their children's length of life, and that the relationship between mother and child was closer than that for father and child (thus contradicting the work of Bell mentioned earlier).

If there is a conclusion that can be drawn from all these studies it is that long-lived parents are not a guarantee of longevity, although they should help one to live longer than offspring from short-lived parents. However, these results are hardly convincing as evidence for a strong genetic involvement in longevity and, if anything, they imply that environmental factors are equally, if not more, important. This likelihood is increased when one considers the fact that both parents and children are likely to have shared similar types of lifestyle, eating habits and social class. In addition, such factors will also have an effect on patterns of disease with life-shortening effects. In summing up much of his work, Louis Dublin wrote in 1949:

> *It may be well, as has been suggested, to seek advantages in longevity by being careful in the choice of one's grandparents, but the method is not very practicable. It is simpler and more effective to adapt the environment more closely to man.*

In other words, having short-lived parents may not be so bad after all, because if we choose to be careful we can probably do a great deal to reduce the disadvantages imposed upon us by their genetic influence.

TWIN STUDIES

The study of twins provides an alternative way of assessing the effects of genetic factors on longevity. Identical or monozygotic twins (accounting for about 1 in 250 births), are created when a single egg (zygote) is fertilised by a single sperm. For some reason the egg then splits into two extra halves giving rise to two fetuses which have exactly the same genetic inheritance. When these children are born they are completely identical and often cannot be told apart. In contrast, fraternal or dizygotic twins (accounting for about 1 in 125 births) originate from two separate eggs (and consequently two different sperms) and therefore result in two fetuses with quite different genetic make ups. These individuals are genetically similar to their other brothers and sisters in that they all share around 50% of their genes. Twins have long been used as experimental subjects in psychology and medicine to test the relative influence of heredity and environment on a wide range of behaviours and diseases. Differences that develop between identical twins must logically be the result of environmental effects since both have the same set of genes, whereas the differences between fraternal twins must be the result of both environmental and genetic influences. The study of twins therefore can help us understand the genetic influence on aging. If genetic inheritance is an important determinant of human longevity then we would predict that pairs of identical twins will have more similar life spans than pairs of non-identical twins.

The most thorough analysis of the aging of twins still remains the work undertaken by Franz Kallmann and his colleagues at Columbia University during the 1950s. These researchers collected the life histories of over 1,700 pairs of twins, of whom approximately a third were identical. Regardless of whether they lived in similar or different types of environments, Kallmann found on average a difference of 36 months (3 years) between the time of death of identical twin pairs. In contrast, fraternal twins of the same sex, on average, died within 74 months (6.2 years) of each other; and this increased to 106 months (8.8 years) for different sex fraternal twins. Not only were identical twins more likely to die of similar causes than fraternal twins, but Kallmann also found that they were remarkably similar in the ways they aged in terms of physical features and mental abilities. For example, he showed that monozygotic twins often had very striking facial and physical similarities even at advanced ages, and the decline of intellectual performance that accompanied aging was much more similar in identical twins than in fraternal pairs.

Surprisingly, there have been few other large scale studies looking at relationship between twins and longevity. The one exception is a Danish investigation undertaken by M. Hauge and B. Harvald that looked at over 200 pairs of monozygotic twins and 400 pairs of dizygotic twins that were born in Denmark between 1879 and 1910. Again, the results showed that identical twins had more similar life spans than dizygotic twins (14.5 and 18.7 years respectively) although the relative difference between the twin pairs was far greater than that found by Kallmann. One reason for this difference is probably linked to the fact that Kallmann only compared twins who had reached the age of 60 years (which ruled out premature deaths) whereas Hauge and Harvald included all ages at which death occurred.

Although these twin studies indicate that genetic inheritance plays an important role in determining life span, there are nevertheless a number of problems associated with interpreting this type of research. For example, most twins share very similar environments and this is particularly the case during their early formative development when their physical resemblance to each other may make parents, and others, treat them as a pair rather than individuals. Ideally, for scientific purposes, it would be much better to compare twins that have been adopted into different environments at birth. Although it is extremely rare, some examples of such twin pairs are known. And in some cases the similarities can be remarkable. For example, Thomas Bouchard working at the University of Minnesota has come across a pair of twins that were separated at birth (both coincidentally named James by their adoptive parents) and who were reunited at the age of 39 years. In terms of physical appearance the two men looked quite different both having different hair styles and clothes. But in terms of lifestyle and previous history the similarities were very close. For example, both drove Chevrolets and enjoyed stock car racing. Both had a background in police work and worked part time as deputy sheriffs. Both chain smoked and enjoyed taking holidays in Florida. Each had built a workshop for themselves in the basement of their house in which one of the twins built miniature picnic tables and the other built miniature rocking chairs. The coincidences even extended to their wives, children and choice of pet. For example, each had been married twice with both of their first wives being called Linda, and both of their second wives being called Betty. The twins also gave identical names to their sons (James Alan and James Allan respectively) and both had dogs named Toy. As perhaps might be predicted, both had similar medical histories: each twin having identical pulse and blood pressure as well as haemorrhoids! Their sleep patterns were also very similar and both had inexplicably put on 10 pounds at the same time earlier in their lives.

Bouchard has now collected information on over one hundred twin pairs (and some triplets) that were separated at early stages in their lives and found many other examples of extraordinary coincidences.

Franz Kallmann provided life histories of some of his twin pairs and these also showed a considerable degree of similarity between identical twins in terms of aging. In one case, a pair of aged identical twin sisters became totally blind and deaf in the same month, and both suffered a massive cerebral haemorrhage on the same day at the age of 69. Despite this, both had experienced very different life histories with one marrying a farmer, raising a large family and living most of her life in the country; whereas the other had remained single, lived in a large city and was employed as a dressmaker.

Despite all these coincidences it must be remembered that not all identical twins show such a high degree of similarity in life pattern or interests. This is not surprising since we are the product of *both* our genes and our environment and this is just as true for aging as it is with many other traits and characteristics. But what are the relative contributions of genetic inheritance and environment to our behaviour including longevity? Geneticists have attempted to answer this question by statistically comparing the differences (or more specifically the 'variance') between identical and fraternal twins on certain behavioural measures. For example, if intelligence is strongly influenced by genetic factors, then each pair of identical twins should have similar IQs; whereas the difference between pairs of fraternal twins (who share 50% of their genes) will be far wider. Given this type of data it is then possible to analyse the variance and arrive at an estimate of the degree of heritability. Using this type of analysis, Bouchard has shown that the heritability of intelligence in his sample of twins is about 70% (i.e., the environment only makes a 30% contribution to intelligence) and that the heritability of personality is about 50%. Of course, all of Bouchard's twin pairs are still alive and so an estimate cannot be made on the influence of genetic inheritance to their longevity. Other studies* have attempted this type of estimate and it appears that the genetic influence on life span is quite low at around 30%. Although it is difficult to estimate the reliability of this figure, it is nevertheless clear that the environment has a very important role to play in determining longevity in humans.

SEX DIFFERENCES

Another possible line of evidence supporting a genetic involvement in aging is the consistently longer average life span of the female

---

*See Table 9.2 in Hartl, D.L. (1991) *Basic Genetics*, page 222.

compared to the male. This relationship appears to hold for animals as diverse as fish, insects, spiders, and mammals. For example, under normal conditions, the female fruit fly will live on average a total of 33 days compared to the male which lives 31 days. Similarly, the female rat typically lives about 900 days whereas the male rat has a life span of approximately 750 days. For the human species, women survive men by a margin of 4 to 10 years throughout the industrialised world. In fact, in all but a few countries (Liberia, Nigeria, Upper Volta, India, Indonesia, Jordan, East Malaysia and New Guinea) the life expectancy at birth for women is found to be consistently higher than for males.

The sole genetic difference between the sexes in humans lies with a single chromosome. We inherit a total of twenty three pairs of chromosomes from our parents, and twenty two of these pairs are identical in males and females. Only one pair of chromosomes show a difference between the sexes, and it is this last pair that determines whether a person is to be a genetic male or female. In short, if we inherit an X and a Y chromosome we become a male, and if we inherit a pair of XX chromosomes we become female. Despite the obvious differences between males and females, it can be seen that genetically the sexes are actually very similar, sharing forty five individual chromosomes and differing in only one! This single chromosome is crucially important, however, because it controls the type of sex hormones that are to be produced in the foetus (and in later life), which in turn determine the development of our reproductive physiology and sex characteristics.

There are a number of reasons why there is a gender gap in terms of longevity. The difference starts at conception with approximately 115 males being conceived for every 100 females. Yet, by birth the male advantage has fallen to 105 because males are more susceptible to spontaneous abortions, miscarriages and stillbirths. Males also have a higher infant mortality rate, and as they pass through adolescence they are much more likely to be involved in fatal or violent accidents, or to commit suicide, than their female counterparts. Consequently, by the age of 30 years the sex ratio is equal, and from then onwards females slowly overtake the males in terms of their numbers. American research has shown that by the age of 65 years approximately 84% of females and 70% of males are still alive; and the reason for the continuing decline in male numbers appears to be due to their much higher susceptibility to lung cancer, pulmonary disease, heart disease, cirrhosis, suicide and accidents.

On first sight these sex differences do not seem to have a strong genetic component. For example, it has often been claimed that society is structured in such a way as to make males lead more strenuous and stressful lives. Because of social expectations and pressures it may be that males are more prone to aggressive and reckless behaviour, suffer more accidents, drink more alcohol and have less time to be interested in their own health. Thus, it is possible to construct environmental explanations for the difference in longevity between males and females. However, it is becoming increasingly clear that this type of explanation does not provide the full story. One of the more provocative studies in this area has been undertaken by Kirby Smith of John Hopkins University who has found a family consisting of four generations in which all the males have the long arm of their Y chromosome missing. The Y chromosome is, of course, the one that determines maleness, and if there is a biological risk attached to being a male, then clearly this is the chromosome which we might expect to be involved in some way. Surprisingly, the initial findings showed that all the males with the damaged chromosome were healthy and normal. However, when Smith compared the average life spans of males and females within the family a different picture emerged. Whereas the females lived on average to an age of 77.4 years, the 14 genetically defect men lived on average to 82.3 years. In other words, they lived nearly 5 years longer! The most obvious conclusion to be drawn from these results is that the Y chromosome contains a gene, or a set of genes, that have an adverse effect on life span. If these genes are only inherited by males then clearly males are at a disadvantage compared to females when it comes to longevity.

Of course, it is difficult to draw any firm conclusions from the study of just one family and consequently we must view these findings with caution. Nevertheless, there is other evidence that supports the premise that the male's genetic endowment does not favour longevity. For example, it is known that up until puberty males and females have similar blood levels of cholesterol, following which they start to increase in the male. The long term consequence of this increase is that males show a marked increase in heart disease after the age of 40 years, whereas this rise does not occur in females until about a decade later after the menopause. The cause of this difference appears to lie with the sex hormones. Male sex hormones such as testosterone increase the blood levels of high density lipoproteins (HDLs), which is the form of cholesterol linked to the narrowing of the arteries, whilst at the same time decreasing levels of the more beneficial low density lipoproteins (LDLs). In contrast, the female sex hormones, such as oestrogen, have the opposite effect by increasing LDLs and decreasing HDLs. However, after the menopause, with the drop in oestrogen (and other female sex hormones), the protective effect of oestrogen is lost.

There are also other biological differences between males and females that may have an important sex-linked genetic basis. For example, one puzzling and unexplained difference between the sexes is that women tend to have more illnesses than men, although their disorders are much less likely to be fatal. One explanation for this difference is that women may have a more efficient immune response than men. In support of this, it has been found that females tend to show more efficient immune reactions including higher antibody responses and increased tumour suppression in a variety of experimental animals, and it is possible that this advantage also holds for humans. It is also known that males tend to be more aggressive than females and that this behaviour is linked to the male sex hormone testosterone. It is also possible that the increase in accidents and suicide found in males (particularly during adolescence when testosterone levels are at their highest) may be partly linked to this hormone and therefore indirectly to the Y chromosome.

SELECTIVE BREEDING OF EXPERIMENTAL ANIMALS

If aging has a genetic basis then it should be possible to modify the life span of animals through selective breeding. One way geneticists have attempted to do this is by looking at the life spans of inbred animals that have been brother-sister mated over many generations. Inbreeding is an important technique in genetic analysis because after several generations inbred lines lose much of their genetic variation and thus individuals become genetically similar. To understand why this occurs it is important to realise that most genes come in more than one form. An example of this is eye colour. The main gene that controls the colour of the pigment of the iris comes in several forms or *alleles*. We inherit two alleles (one from each parent); but what happens if we inherit a blue allele from one parent, and a brown allele from the other? The answer is that we will inherit brown eyes because the brown allele has *dominance* over the *recessive* blue allele. Thus, not only do we have different forms of the same gene (alleles), but they can also be dominant or recessive. If a person inherits a matched pair of alleles (as with blue eyes) he or she is said to be homozygous (from the Greek meaning 'identical') whereas a person inheriting one dominant and one recessive gene is said to be heterozygous (from the Greek for 'different').

Returning to the original question of why inbreeding leads to genetic similarity, the answer is that it increases the chances of producing homozygous genes. And, once homozygous genes are produced in an inbreeding stock they are not easily bred out of successive generations unless new genetic material is introduced. In fact, typically after 20 successive single-pair brother-sister matings, all members of an inbred strain are as genetically alike as they are ever likely to get. Despite this, each strain will have their own unique combination of homozygous genes which, of course, means that they will be genetically different to other inbred strains.

One of the first studies to examine the effects of inbreeding on longevity was undertaken by Pearl and Parker in 1922, who selectively bred the fruit fly (*Drosophilia melanogaster*). Through many generations of inbreeding, they managed to produce 5 different strains of fly each with its own characteristic average longevity. These ages ranged from 14 days in the shortest-lived strain, to 49 days in the longest-lived strain. Furthermore, once each strain was developed, the flies continued to produce generation after generation of offspring each with its own average life span typical of that strain. These findings clearly show the importance of genetic make-up on longevity. However, despite this, the ages of all the strains obtained by Pearl and Parker were shorter than those found in normally bred fruit flies showing that inbreeding is not an advantageous strategy for longevity. One reason for this situation is that many harmful genes are often 'hidden' because they are normally recessive; but with inbreeding they often find themselves matched with identical genes which makes them become dominant (this also explains why inbreeding in humans is often associated with mental defects and other kinds of abnormality).

Interestingly, crossing two inbred lines (with short life spans) often results in hybrids with significantly longer life spans. For example, Pearl showed that when two lines of fruit flies with age spans of 37 days and 21 days were crossed together, the offspring (or F1 generation) had a life span of 47 days. Thus, the effects of inbreeding can be quickly undone and the most likely explanation for this effect is that the outbreeding results in the increased likelihood of heterozygous genes, with harmful genes reverting back to their recessive status.

Although this type of research is of theoretical interest and shows that the overall genetic constitution may to some extent contribute to shortening life span; the vast majority of populations are, of course, not artificially inbred, and furthermore this approach tells us little about the nature of the genes that are acting to increase life span. If genes controlling longevity really do exist then it should be possible to increase life span by selectively breeding animals that carry these genes (i.e., the offspring from long lived parents). This approach has been adopted by several researchers with considerable success. For example, Michael Rose at the University of California started with a group of fruit flies and began collecting eggs to create new generations from 'old' adults that were 28 days old. After four generations, Rose started to collect the eggs from 35 day old flies, and after a few more generations it was possible to collect eggs from 42 day old flies. In fact, Rose continued this strategy until he was collecting eggs from 70 day old flies. At this point it was clear that the breeding procedure had significantly increased longevity in the offspring of these flies. For example, compared to non selected flies, the late-reproducing flies increased their average life span from 33 to 43 days for females, and from 39 to 44 days in males.

It is clear that these differences in longevity have a genetic basis. Moreover, Rose has shown that whatever these genes are doing, they are in effect acting to delay the onset of senescence. For example, he has found that not only do the longer-lived flies start their breeding cycles later in life, but they also produce relatively more eggs, laying an average lifetime total of 1,733 eggs compared to 1,635 for the control group. Rose also argues that these findings lend support to the antagonistic pleiotropy of aging (see chapter 2) as these late-reproducing flies have effectively traded-in the advantages of early reproduction for increased longevity.

A number of studies have followed up this work by examining the physiological characteristics of purposively bred long-lived flies. It has been found that these flies are significantly more resistant at all ages to starvation, desiccation (i.e., a dry environment), heat stress and alcohol vapour poisoning. In addition, they show lower rates of respiration and flying activity at younger ages along with an increase in fat deposits. These characteristics are likely to be beneficial for longevity since decreased respiration will decrease the rate of metabolism by reducing the amount of food and water consumed (see also chapter 7) while increased fat reserves will help protect them against starvation. Rose and his colleagues have also taken this research one step further by selectively breeding flies that are resistant to the stresses of either starvation or desiccation. Interestingly, starvation resistant flies are not resistant to desiccation, and vice versa, indicating that there is more than one genetic route to extending aging.

HOW DO GENES EXTEND LONGEVITY IN FRUIT FLIES?

The fact that longevity can be selectively bred into fruit flies implies that a genetic mechanism must somehow be involved in controlling the aging process. But how can this occur? Or more specifically, what possible biological functions can these genes be regulating to control longevity? One of the first studies that was brought to bear on this problem was undertaken by Leo Luckinbill at Wayne State University who attempted to identify the chromosomes on which the longevity genes were located. His strategy was to breed strains of genetically identical *Drosophilia* that only varied in terms of a single chromosome (these flies normally have only 3 chromosomes). The results showed that all of the chromosomes examined by Luckinbill were to some extent contributing to longevity, but one chromosome stood out as much more important than the rest. This was chromosome number 3 and it accounted for about two-thirds of the fruit flies' life span. This was a particularly interesting finding since this chromosome in *Drosophilia* is also known to contain the genes that code for the antioxidant enzymes catalase and superoxide dismutase that help protect the cell against highly damaging free radicals caused by the continuous break down of oxygen (see also page 49).

Could it be, therefore, that fruit flies selectively bred for their longevity were living longer because they contained higher levels, or more efficient forms, of these protective enzymes? This question was examined by the American researchers Steven Dudas and Robert Arking who bred a strain of long-lived fruit flies and then measured the expression of genes controlling the production of antioxidant enzymes. The results showed that the longer lived flies did indeed show higher levels of these protective substances. Moreover, this increase in antioxidant activity began early in life and lasted about 3 weeks which was a period in which the long-lived flies were less susceptible to the effects of free radicals (as measured by exposure to paraquat) than controls. These results indicate that one of the reasons why artificially bred long-lived flies live longer is because of gene combinations that result in more efficient free radical protection than that found in normally bred flies.

But perhaps the most convincing work of all, showing the importance of antioxidant enzymes in fruit fly longevity has been performed by William Orr and Rajinda Sohal working at the Southern Methodist University in Dallas. These researchers have gone one important step further and created new strains of transgenic fruit flies that carry extra copies of the genes known to regulate the activity of the antioxidant enzymes. Transgenic animals are reared from embryos into which new DNA has been introduced by genetic manipulation of a fertilised egg or very early embryo outside the mother. If the new DNA becomes strongly incorporated into the chromosome at this early stage, it will then be present in a large number, if not all of the cells, of the developing embryo and later adult. Transgenic techniques have been applied to several laboratory animals (particularly mice) and several domestic animals including pigs, sheep, goats and cows. Orr and Sohal applied a similar technique to flies and implanted three extra copies of the genes responsible for the antioxidants superoxide dismutase and catalase. The results showed that these transgenic flies exhibited a one-third extension of life span, increased physical performance, and reduced amounts of protein damage caused by oxidative stress compared to normal flies.

The evidence therefore appears to be reasonably clear-cut. Increased levels of antioxidant enzymes increase the life span of fruit flies, thus supporting the theory that oxygen mediated damage is an important cause of aging. But what about other species? Although it is probable that the same findings will extend to other types of animal, such work has yet to be undertaken, and we can only wait and see whether this will prove to be the case. As Michael Rose points out in a *Scientific American* article published in December 1992, the work undertaken on fruit flies can be seen as a trial run for testing the same hypothesis with more advanced species such as laboratory mice. In short, if long-lived mice can be created, and the specific genes responsible for increasing longevity identified, then these discoveries should lead to a much closer approximation of what causes aging in humans. However, such work does not come cheap. Rose estimates that such long term studies with mice alone would cost in the region of $10 million and require considerable planning and organisation.

MUTATIONS AND AGING

Fruit flies are not the only creature to have given us insights into the genetic basis of aging. One of the most useful organisms of all is the microscopic nematode worm (*Caenorhabditis elegans*). There are in fact over 12,000 known species of nematode including a few that are found as parasites in humans including pinworm, hookworm and thread-worms. It is the soil nematode *C elegans*, however, that has attracted the attention of genetic researchers. Firstly, these worms (which are about 1 mm long) can be cultured like bacteria in dishes and grown easily under laboratory conditions. Secondly, they contain a relatively simple genetic system consisting of 6 small pairs of chromosomes that hold around 3,000 genes, most of which have been identified and mapped out. Thirdly, and perhaps most importantly, *C elegans* only contains only 959 body (soma) cells which can easily be seen because the nematode has a transparent body. This important characteristic has allowed researchers to follow and to map out the fate of every cell at every stage of the animal's development from fertilised egg to adult. Consequently, it is fair to say that more is known about the genetics and development of this tiny creature than any other living organism.

One of the great benefits of using nematodes is that there is a relatively simple way of determining the function of their genes. This involves causing genetic mutations by exposing batches of nematodes to toxic chemicals or radiation. Large numbers of nematodes are often killed or injured by such treatment but occasionally a genetic mutation is caused which produces a clearly definable biological effect. By examining the consequences of such a mutation a greater understanding of the gene's normal function can be worked-out. This approach may at first sight appear to be an unpromising one for aging research since mutations might be expected to decrease rather than to increase longevity (although surprisingly this rarely occurs in the nematode). However, in 1983 a mutation was produced that was actually found to cause an increase in the nematode's life span. This mutant gene (which was called *age-1*) has been extensively studied by Thomas Johnson working at the University of Colorado who has shown that it extends nematode longevity from an average of 20 days, to about 34 days – that is, an increase of 65%. The question that immediately followed this discovery was: how could such an effect possibly occur?

One of the most interesting aspects of the *age-1* gene is that although it causes no change in the length of the nematode's period of sexual maturity, the gene nevertheless results in the nematode producing significantly fewer offspring. For example, in one study, *age-1* mutated nematodes produced 75% fewer eggs (72 versus 317) compared to normal worms. It was almost as if 'energy' normally used in reproduction was somehow being shifted away into promoting life extension. But what biological form could this 'energy' be taking? Again, the answer seems to be through increased free radical protection since it has also been shown that *age-1* mutant nematodes enjoy elevated levels of the antioxidant enzymes superoxide dismutase and catalase. Thus, it appears that the *age-1* gene does not increase longevity by stretching out the nematode's developmental stages (in fact it undergoes 4 separate larval stages before becoming a sexually mature adult); but rather the *age-1* mutation provides the nematode with added protection against the adverse effects of aging, which according to Johnson acts to increases the nematode's life span by extending the period of its post-reproductive decline.

More recently, however, the increased life span of the nematode has been further extended by other types of mutations. For example, in 1996, Bernard Lakowski and Siefried Hekimi of McGill University in Montreal, identified a set of 4 new genes (which they call *clock* genes) that were found to increase the nematode's life span by about 50%. Remarkably, when these researchers bred worms carrying both types of mutation (*age-1* and *clock*) they actually managed to produce a strain that lived up to two months (some 5 to 6 times longer than normal!). Perhaps even more important was the finding that nematodes with *clock* mutations appeared to show very different patterns of aging than those animals with *age-1* mutations. This pattern is perhaps best described as a stretching-out or slowing-down of life span. For example, cells taken from *clock* nematodes take longer to divide in the embryo, their larvae take more time to turn into adults, and even the rhythmical thrashing movement of the nematode when it is placed in water is slower than normal. In short, the *clock* genes seem to be slowing down the speed at which the animal lives its life, presumably by slowing down the speed of its metabolism in some way. These findings also appear to lend support to the rate of living theory which will be discussed in more detail in chapter 7.

This slowing down of life span can also be explained in terms of free radical chemistry since if energy (oxygen) is used-up more slowly in metabolism, then fewer free radicals will be produced and less oxidative damage will occur. Both the *age-1* and the *clock* mutations, therefore, may be increasing longevity by reducing the effects of free radical damage. That is, the *age-1* mutation may be promoting longevity by increasing levels of superoxide dismutase and catalase that protect against free radical damage; whereas the *clock* mutation may be promoting longevity by slowing down the rate at which free radical reactions occur. Clearly, if there is a single conclusion we can draw from all this work, it is that free radical reactions undoubtedly play a major role in the process of aging. This point is also touched upon at several other places in this book.

Genetic Diseases with Special Relevance to Aging

Although there are many advantages of using nematodes to examine how genes can affect the fundamental processes of aging, this research unfortunately tells us little about the role of specific genes in determining human longevity. One way of overcoming this problem is to examine the effects of human genetic disease on aging. In other words, if there are specific genes involved in human aging, then we might expect that they will be exposed by certain types of genetic disorder. Indeed, there are nearly 7,000 known human inherited conditions to choose from, although none appear to increase longevity. However, a number of these disorders reduce life expectancy, and more importantly, there are a few that actually appear to result in accelerated or premature aging. George Martin, working at the University of Washington, has identified at least 10 genetic disorders (which he calls "segmental progeroid syndromes") that result in premature age-related pathology. Interestingly, these conditions differ in their patterns of aging and none show all the common features of senescence, indicating that there are many genes, possibly scattered over several chromosomes, that have some type of effect on human aging. The strong suspicion exists (although not all researchers agree) that if only we can understand the nature of these diseases, we can do much to help explain the genetic basis of human aging.

PROGERIA

Progeria, also known as Hutchinson-Gilford syndrome (named after two British doctors who independently described the condition in the late 19th Century) stands out as the most striking of all the diseases of premature aging. This disease tragically affects young children who often resemble little old men and women by the time they reach early adolescence. Although the progeric child may appear to be normal in its first year, it is not long after that a severe growth retardation begins to manifest itself, and is often accompanied by balding with loss of eyebrows and eye lashes. By the age of 5 years the child often looks old and frail with grey thinning hair and wrinkly skin, and there is often a loss of fatty tissue under the skin giving it a transparent appearance that makes the blood vessels (especially those over the scalp) stand out. The face is particularly distinctive with prominent eyes and a hooked nose giving the appearance of a 'plucked' bird. The bones if they grow at all (a teenage progeric is typically no taller than 40 inches tall and weighs around 25 to 30lb) are weak and easily dislocated. Consequently, there is often a peculiar bow-legged or 'horse riding' posture along with stiffening of the joints, particularly of the hands and fingers. Sexual maturation does not occur and the voice remains weak and high-pitched.

Progeric children do not live very long. The average age of death is 12 years, although sometimes they can live until their early 20s (the maximum age recorded is 27) and over 80% of deaths are due to heart attacks or congestive heart failure. In fact, a progeric child may have clear signs of heart disease by 5 years of age with narrowing of the arteries, elevated cholesterol, high blood pressure, and angina pectoris. At post-mortem there is often advanced atherosclerosis and fibrosis of the heart similar to that found in old people with long term vascular disease. However, progeria does not mimic aging in all respects. For example, progeric children do not show malignant tumours, arthritis, diabetes, cataracts or deafness, all of which are common features of old age. Furthermore, mental impairment and senility are not observed, and if anything, progeric children show normal to above average intelligence and are often painfully aware of their condition. At post-mortem there are no signs of brain atrophy or age pigment (lipofuscin) accumulation that is also frequently associated with old age.

Fortunately, progeria is an extremely rare condition. It has been estimated that the disease occurs in about one in eight million births, and in 1972 a survey found only 60 cases world-wide. The condition appears to inflict males and females equally and is probably found throughout the world. Its rarity makes the disease difficult to investigate and, because of this, the actual cause of progeria is not known for certain. The general consensus is that progeria arises spontaneously from a dominant mutation in one of the germ-line chromosomes although the nature of genes involved are not known. It has been suggested that the mutated gene(s) may act to inhibit infant growth and development, and possibly prevents the production of a further chemical factor that normally switches on other genes involved in the developmental process (such as those involved in sexual development). Unfortunately, this is mere speculation: until the sites of the damaged genes are tracked down, and their functions established, the disease and its true relationship with aging will remain controversial.

WERNER'S SYNDROME

Werner's syndrome is sometimes called 'progeria of the adult' and was first described by Werner in his doctoral thesis in 1904. Victims of this disorder generally first start to show symptoms of premature aging around the age of 20 years of age, although in some cases the first manifestations may not begin until 40. The first sign is nearly always greying or whitening of the hair, followed by changes in the skin which shows patches of dermatitis and blistering (particularly of the ankles) accompanied by general wasting of body's muscles. Early formation of cataracts leading to poor vision and partial blindness are common features and there is also an increased likelihood of diabetes mellitus. As with progeria, the most common form of death is from cardiovascular disease and stroke although approximately 10% of victims also develop cancers with a high proportion of sarcomas (cancer of connective tissue) and meningiomas (brain tumours). By the age of 35 or 40 years, a person with Werner's syndrome usually has a short bent stature (although not as short as progeria) with baldness or thinning grey hair, poor vision, muscle wasting and osteoporosis. All of these changes are signs of aging that are normally only found in people who are at least double their age. However, as with progeria, there does not appear to be any intellectual impairment or dementia, and the victim is fully conscious of their condition.

Werner's syndrome is about twice as common as progeria though this still makes it a very rare condition. Although victims reach sexual maturity and are capable of reproducing, the foetus is typically aborted, or dies prematurely, so that the genetic defect is not passed on to future generations. The genetic mode of inheritance appears to be very different to progeria since Werner's syndrome is believed to be an autosomal recessive condition which means that the victims must inherit a copy of the mutant gene from *both* parents. Surprisingly, it has been estimated that 1–5 per thousand people may actually carry the defective gene although the chances of two people bringing it together to produce a child with Werner's syndrome is about 1–25 per million. In 1992 the location of the mutant gene was narrowed down to chromosome 8 by the Japanese researcher Makoto Goto who examined the DNA of 21 families containing 31 affected individuals. It is possible that progeria may have a similar chromosomal location although this has not yet been shown. At the time of writing the actual gene has not been isolated, although it should be only a matter of time before it is identified and work can take place on what the gene is actually doing to produce the symptoms of the syndrome.

DOWN'S SYNDROME

According to George Martin the disease that shows more signs of accelerated aging than any other is Down's syndrome which occurs in about 1 in 700 live births. Children with this disorder typically have a distinctive sloping face with slanting eyes which originally gave rise to the unfortunate term of mongolism. They also tend to have a short stature accompanied by short thick hands and feet, and the vast majority show some degree of mental handicap. However, less well known is the fact that as they grow older they also tend to demonstrate signs of accelerated aging, and have a life expectancy of around 48 years (in fact before the invention of antibiotics this figure was nearer 20). These signs includes greying and loss of the hair, increased incidence of malignant cancer including leukaemia, cataracts, cardiovascular degeneration, increased age pigmentation, and increased autoimmuninity (see also chapter 10). In addition, individuals with Down's syndrome (unlike those with progeria and Werner's syndrome) undergo marked mental decline, and at post-mortem reveal many of the histological changes of the brain commonly found in old people with senile dementia including those with Alzheimer's disease.

The genetic basis of Down's syndrome is well established. It is not caused by a mutation as such, but is due to the cells of the body containing an extra copy of chromosome 21. In other words, there are three copies of chromosome 21 instead of the normal pairing of two, and for some reason this extra chromosome adds an 'extra dose' or interferes with the way in which all the genes are expressed. So far 16 genes have been traced to this chromosome including importantly the one that codes for the enzyme superoxide dismutase, which as previously mentioned, helps protects the body against free radical damage. It would perhaps be surprising if this particular gene was not found to be somehow involved in the premature aging associated with Down's syndrome, although as we shall see below, there are also other genes on this chromosome that are probably linked to the aging process.

ALZHEIMER'S DISEASE

Alzheimer's disease is a form of dementia that is increasingly taking its toll as more and more people live to longer ages. It has been estimated that 5% of people over the age of 65 years, and 80% of people over the age of 80 years suffer from the disease, and there are probably some 500,000 to 700,000 victims in the UK alone. In the USA it has been cited as the fourth most common form of death in the aged although it is rarely recorded as such on death certification (the final cause of death is more likely to be stated as pneumonia, infection or asphyxiation). The age group most clearly at risk are the over-eighties, who also happen to be the fastest growing segment of the population in most, if not all of the Western world; and this inevitably means that the incidence of Alzheimer's will steadily increase over the next decade or so. It is perhaps not surprising therefore that Alzheimer's disease has been called the disease of the century. The disease often progresses rapidly, transforming a normal healthy person into a living vegetable within 5–7 years, destroying memory, personality and even one's sense of identity. Dr Jonathan Miller once poignantly pointed out that by the end of the disease, one is effectively left with a corpse that the undertaker has forgotten to collect.

Despite its obvious symptoms, Alzheimer's disease can only be diagnosed with absolute certainty at post-mortem. Frequently, although it is not always the case, the most striking change is marked atrophy and degeneration of brain tissue, particularly of the cerebral cortex. It is a sobering fact that in adulthood we lose a small number of irreplaceable brain cells every day of our life (possibly as many as 100,000 each day), and by the age of 80 a 'normal' brain will have decreased in weight by some 10–15%. However, in Alzheimer's disease there may be over a 30% loss of brain tissue and this often in much younger individuals. Histological and microscopic examination also shows a number of abnormal features including the scattered deposition of a hard protein substance called amyloid (meaning 'starch-like') that accumulates in the blood vessels and forms small hard plaques in the brain tissue. Another histological indicator of Alzheimer's disease are small tangles of filaments known as neurofibrillary tangles that were first identified by Alzheimer himself in 1908. There is also a marked loss of certain brain chemicals, in particular the enzymes necessary for the synthesis of the neurotransmitter acetylcholine. In general, the more severe and advanced the illness, the more pronounced these degenerative changes.

Until fairly recently, most cases of Alzheimer disease were not generally regarded to have a genetic origin, although a very rare genetic form of the disorder was known. This type of Alzheimer's disease (sometimes called familial Alzheimer's disease) had been found to be confined to a few (typically large) families, and examination of inheritance patterns showed that the disease was being transmitted in an autosomal dominant fashion (that is, if one was unlucky enough to inherit the gene, then one was destined to develop the disease). This form of Alzheimer's disease was known to have an early onset (between 40–60 years of age) and a very severe progression. In 1987, the site for this familial Alzheimer's gene was found to be located on chromo-some 21 (the same chromosome that is linked to Down's syndrome). This generated a great deal of interest since older individuals with Down's syndrome also show many of the histological features associated with Alzheimer's disease including the deposition of amyloid, neuritis plaques and neurofibrillary tangles. Despite this, the abnormal gene was only present in a small number of Alzheimer cases, and the vast majority of people with the disease did not carry this gene.

In 1992, a second genetic site was identified on chromosome 14, although again this was only found in a few individuals and was associated with early onset forms of Alzheimer's disease. This finding raised the possibility that the disease might be the result of several genes acting together and not due to a single gene. In one sense this prediction has turned out to be true. In 1993, a site on chromosome 19 was also found to be linked to Alzheimer's disease although this time there was a big difference. This gene was not a mutation and moreover it came in three forms (or alleles). The gene was called the APO gene (because it was known to be involved in the creation of a substance called apolipoprotein found in brain cells) and the three alleles were designated APO-2, APO-3 and APO-4. It soon become clear that everybody carries two copies of this gene, one from each of their parents, and consequently many different combinations of the genes are possible (see table 3.1). The startling fact that arose from this discovery, however, was that the inherited combination of genes was linked to Alzheimer's disease in both its early and late onset forms. The worst combination of genes to inherit are APO-4 from both parents. Inheritance of this combination shows that the average age of Alzheimer's onset is before the age of 70 years, (although fortunately it appears that only 2% of the population contains this 4/4 genotype). The most common genotype is in fact the 3/3 combination which predisposes its carriers to Alzheimer's disease around the ages of 80–90 years (60% of the population carry this genotype). The most favourable combination of alleles, however, is the 2/2 combination. Although this pair of alleles are found in less than 1% of the population, this combination seems to provide considerable protection against Alzheimer's possibly up to, and beyond a hundred years. But, despite this, the implication

Table 3.1 Mean ages of onset of Alzheimer's disease as a function of the inheritance of APO genotypes (data from Roses, A.D. (1995) *Scientific American*, Sept/Oct, 16–25)

Genotype	% U.S. Population	Average age of onset	Range
2/2	<1	?	?
2/3	11	>90 years	50–140
2/4	5	80–90 years	50–>100
3/3	60	80–90 years	50–>100
3/4	21	70–80 years	50–>100
4/4	2	<70 years	50–>100

seems to be: if we live long enough, all of us will eventually develop Alzheimer's disease.

Considerable research is presently underway to try to establish what apolipoprotein does in the brain, and understand why the combination of APO alleles makes such an important difference to its function. The issue of Alzheimer's disease also raises some difficult questions concerning the nature of aging itself. Most people regard Alzheimer's disease first and foremost as a disease, and different to aging which (as discussed in chapter 1) is generally regarded as a natural loss of function that occurs with time. However, it appears that the incidence of Alzheimer's disease also increases with time and it involves a gene combination that appears to be a normal part of our inheritance. Thus, in this particular case, it is difficult to draw a clear distinction between normal and pathological aging. Clearly, the APO genes do have a bearing on life span, and an understanding of the permutations of how these genes interact and manifest themselves may not only lead to greater insight into how genes control the development of aging, but also help define more precisely what aging is.

SUMMARY

- Genes can be viewed in two ways. Firstly, they are units of inheritance that are passed from parents to offspring. Secondly, genes exert important control over the biochemical activity of the cell.
- Life span is the result of both genetic and lifestyle factors. It is difficult to gauge the relative influence of each, but for most people the lifestyle factor is probably the more important.
- Genealogical studies show that long-lived parents tend to live longer than their offspring and that short-lived parents tend to produce offspring that are longer-lived.
- Twin studies have shown that, on average, identical twins have more similar life spans than non-identical twins.
- Throughout the animal kingdom it is generally found that females outlive males. In the case of humans, females tend to outlive males by a margin of 4 to 10 years.
- The inbreeding of animal populations tends to reduce genetic variation by producing genes that are homozygous which is usually disadvantageous for longevity.
- Selective breeding of long-lived animals has been shown to extend life span. Michael Rose has shown that fruit flies bred in this way start their breeding cycles later in life, but in total produce more eggs.

- Transgenic flies with extra copies of the genes that encode for anti-oxidant enzymes (which protect against free radical damage) have their life span increased by about one-third.
- A number of gene mutations in the nematode worm have been shown to increase longevity. These gene mutations appear to work in one of two ways: they either increase free radical protection or they slow down metabolism.
- Progeria, Werner's syndrome, Down's syndrome and Alzheimer's disease all show certain signs of premature or accelerated aging. The genes involved in these disorders may (or may not) be connected with 'normal' aging.

GENETIC THEORIES OF AGING

Theories of aging can be divided into two broad groups: those that presume a preexisting master plan, and those based on random events

–**Leonard Hayflick**

Two out of every three people reading these words will die for reasons connected with the genes they carry.

–**Steve Jones**

Within the nucleus of nearly every cell in our body lie 23 pairs of chromosomes which contain the genes that we inherited from our parents at the moment of our conception. Not only do our genes provide the blueprint by which the body is made, but they also continue to function throughout life to make a variety of biochemicals (mainly proteins) that are essential for life. As we have seen in previous chapters, there is overwhelming evidence that some genes underpin the aging process. Indeed, research shows that there are genes which make antioxidant enzymes that protect us against free radical damage, and others that might be controlling the rate of metabolism. But one question remains unanswered: what happens to these genes (and others) over the life span? In short, there are three main possibilities: (1) our genes slowly accumulate damage and therefore become defective; (2) they are programed to cause aging and senescence; or (3) they continue to work perfectly throughout life but are forced to look on helplessly as our bodies succumb to other types of damage. Since all biological systems appear to show some degree of age-related decline (and why should our genes be any different?) then the last option is perhaps the least likely. Even if it was true, it does not represent a genetic theory of aging and so will not need to be discussed any further in this chapter. Thus, if we accept that genes have a role to play in aging, this leaves us with two options, and brings us to one of the most crucial questions of all: is aging due to the gradual accumulation of accidental genetic damage (this type of process is often called 'stochastic' meaning random and occurring by chance) that results in biological decline; or is genetic change underlying aging a result of a clock-like programed process that is preplanned and deliberately meant to occur? This is perhaps the one question, above all others, whose solution may provide the most important key to unlocking the mysteries of aging.

A CLOSER LOOK AT DNA

We have already had a look at the DNA molecule in chapter 2. As we have seen, it consists of two long spiralling backbones (made of phosphate and sugar) wrapped around each other like two strings of a rope, and held together by pairs of complementary bases (labelled A,T,G and C). Although this genetic alphabet is very simple and composed of only 4 bases, there are astronomical numbers of bases on each stretch of DNA (or chromosome) making the content of the overall message incredibly complex. It has been estimated that the human genome (i.e., the total compliment of all 23 pairs of chromosomes) contains over 3,000 million base pairs, and this breaks down into approximately 6.5 million base pairs, (or paired 'letters') for each individual chromosome. To put this figure into perspective, the 1969 edition of the *Encyclopaedia Britannica* (which also coincidentally consists of 23 volumes), contains only 200 million letters; or put another way, enough information to fill approximately three chromosomes. The evolution and packaging of chromosomes and all its DNA into the cells nucleus, however, is perhaps the most staggering achievement of all. To enable over 3,000 million base pairs to exist together, each individual cell of the body contains about 6 feet of DNA crammed into a nucleus that is about 0.005 mm in diameter. This is a remarkable feat of biological engineering by any standards, although the complexity does not end there. As Steve Jones has pointed out in his book *The Language of the Genes,* a useless but amusing fact is that if all the DNA in every cell of a human body was stretched out it would reach to the moon and back 8 thousand times! Put another way, if we could somehow harvest all the genetic information from all our cells, it would be equivalent to the amount of information in the *Encyclopaedia Britannica* many trillions and trillions of times over. This is what it takes to produce and maintain life in something as complex as a human being.

From the moment of conception to the moment of death, our genes (made up of base sequences) continue to function to provide the biochemical processes necessary for life. The vast majority of genes make proteins, which in turn serve a huge range of biological roles (e.g., they form the structural components of our cells and tissues, and make the enzymes necessary for their function). Proteins are unique: they are large (in a molecular sense) and structurally very complex (their shape is particularly crucial to their function). If one could dismantle a protein, one would find that they are comprised of long chains of amino acids (the simplest proteins contain only one chain) which are typically bent, pleated and twisted at various places to create interlocking 3-D shapes. Each gene is responsible for making a protein; although it is perhaps more accurate to say that genes are a code for the correct order of amino acids to be put together, that in turn make a protein (the actual construction of a protein takes place at special sites called ribosomes away from the nucleus and this will be discussed in the next chapter). Despite the great complexity of our genes, they are only capable of making 20 types of amino acid, although from this limited pool a large variety of proteins can be constructed. In fact, the creation of each amino acid is derived simply from a code of just 3 consecutive DNA bases (such as CGA or TGG) otherwise known as codons. In short, a single gene is basically a long sequence of codons that code for all the amino acids necessary to assemble a specific protein.

It has been estimated that our chromosomes contain around 100,000 genes (maybe a few less), with most specifying the design of a specific protein (although a few genes make other types of molecules such as RNA). However, one of the most surprising aspects of our DNA is that less than 5% of our genome appears to be composed of genes, and the rest (sometimes called 'junk' DNA) seems to be largely redundant. In fact, if we could somehow enter the genetic world, and travel along our chromosomes, we would find that most of our DNA is composed of endless repeats of the same 'nonsense' message, often in the form of 5 or 6 bases repeated next to each other (such as ACCTGACCTG). These bases may also be interspersed by simpler repeated sequences, such as the two bases C and A, that can be multiplied many thousands of times. At other locations along the chromosome there may exist long and complicated sequences whose message, if meaningful, remains unknown. Thus, the terrain of the chromosome that holds the secrets of life is a surprising and often bizarre world.

After long stretches of seemingly nonsense DNA, however, we eventually come across sequences of meaningful codons of 3 letters that are known to code for amino acids (at which point it is probable that we have discovered the start of a gene and this would be confirmed if we also find many chemical reactions taking place in the vicinity). These 'working' stretches of DNA come in many different shapes and sizes, with the smallest gene consisting of around 500 bases, and the largest containing well over 2 million. However, the codons that compose the functional part of the gene are not as tightly clustered as we might imagine since they are frequently interrupted by long lengths of non-coding DNA called introns. In fact, large genes may contain over 40 intron spaces with over 100 times more DNA than the working part of the gene. One such large gene is the one that causes haemophilia. In its normal state this gene makes a protein that allows the blood to clot in response to injury, but in people with haemophilia this gene has a mutation that results in it producing a faulty protein. Remarkably, this protein is constructed of some 2,332 amino acids, yet 95% of the gene appears to be made from introns that do not contribute to the gene's output! This may appear to be a strange way of going about things, but for whatever reason, this is the way DNA has evolved.

Although we have around 100,000 genes scattered throughout our chromosomes, most of these in any given cell are turned off, which means that only a fraction of our genes are working at any one time. To understand why, it is important to remember that nearly every one of the 60 trillion cells in our body contain the same genetic material, despite the fact that cells come in many shapes and sizes, and perform many different functions. The important question is, therefore: why do some cells become nerve cells whilst others turn into liver, skin, heart and so on? Without going into great detail, the simple answer is that nerve cells express a different pattern of gene activity than say, heart cells. Thus, each cell in the body has the overall blueprint to make the whole body, but of course, what happens in practice is that the cell only uses the genes that are relevant to its own function. However, this raises another important point about our genetic world – there must also be executive genes that somehow decide what other genes are to be switched on, and off. Thus, it appears that some genes are more important than others.

Finally, a simple question: why is it that all human beings share the same genes, yet turn out to be individually different (with the exception of identical twins)? The answer is that we all have different alleles; that is, genes that have evolved into slightly different forms (see page 64). We all have differing sequences of bases within our DNA, and on average two people will differ in about one base every thousand which means that there are about 3 million differences in the genetic message between any two people. This provides the basis of genetic individuality. Put another way, all our genes still serve the same function but their expression may be modified slightly by base structure changes (as in the case of eye colour). These changes have come about through chance alterations or mutations that have taken place in the structure of the bases throughout evolution. The consequences of a newly formed base alteration, whether by the simple addition or removal of a base, will depend on where it occurs in the DNA. If a change occurs in the non-functional (intron) part of a gene then there will probably be no effect on the individual. But, if a base change takes place in an important codon that encodes for a crucial amino acid then the consequences may be severe resulting in the development of genetic disease. However, not all base alterations will have harmful consequences, and some will alter the functioning of the gene only slightly thereby producing alleles with no adverse effects to the individual. Multiple alleles have been found for about half of the human genes so far examined; and it has been estimated that the number of potential allele combinations is so vast, that everyone alive today is genetically different to each other, and to every person that has ever lived. But perhaps even more important is the fact that this example shows that DNA has the capability to change over evolution. In fact, as we shall see below, the structure of our DNA is also subject to change across our life span, and this may also be an important cause of aging.

GENETIC THEORIES OF AGING

Many of the secrets of aging probably lie within this strange world of DNA. But to show conclusively that such a genetic link exists is far from easy. There are two main obstacles: (1) it must be shown that DNA is either susceptible to damage that occurs with time, or that there is a programed unfolding of genetic change over the life span; and (2) it must be shown that the outcome of such genetic change is the functional decline of the organs and tissues of the body that we normally associate with aging. It is perhaps not surprising that the genetic theory of aging still remains largely unproven experimentally although there are no shortage of theories. We will begin by looking at three theories (somatic mutation, DNA repair and redundancy theories) that all view aging as arising from wear and tear, or random genetic decline.

THE SOMATIC MUTATION THEORY

Several genetic diseases have already been mentioned in this book (e.g., haemophilia, Progeria and Werner's syndrome) that arise from structural changes, or mutations, occurring in certain genes. To be more precise, all of these diseases are the result of mutations that occur, or are present, in the original germ cells. Obviously such damage will have profound consequences because the fertilised egg will carry the faulty gene, and thereafter so will every somatic cell of the body. Indeed, mutations that occur in the germ cells often have a lethal or damaging effect on the individual. Of course, mutations can also arise in the DNA of somatic cells after the individual has been fully formed (which are called 'somatic mutations'), although in this instance the consequences are likely to be much less severe. The main reason is that there are millions of somatic cells, even in the smallest body organ, and if a few individual cells start to carry damaged genes the overall net effect is not likely to be life threatening.

This is not to say, however, that somatic mutations are unimportant. On the contrary, the number of somatic mutations will probably increase steadily during life, and if so, the cumulative effects may well eventually have important consequences. This is essentially the somatic mutation theory of aging – a theory that has remained central to genetic explanations of aging for the last 40 years or more. This theory rests upon the assumption that, over time, somatic cells will become increasingly susceptible to the cumulative effects of DNA damage, and this will result in a number of changes including cell loss, the build up of abnormal cells, and the development of impaired cells that are no longer able to fulfil their proper functions. In short, these are the types of change that are believed to lead to biological decline, and there are many ways in which this could happen. For example, if we examine the heart, it has been suggested (although not proven) that mutations in the cells of the smooth muscle wall of the heart's aorta may lead to the formation of atherosclerotic plaques. Alternatively, somatic mutations may lead to a loss of heart cells, or cells not responding correctly to chemical signals from the rest of the body. These are just some of the examples from a myriad of possibilities that show how somatic mutations might lie behind the aging process.

If the number of somatic mutations do accumulate with age then we would expect to find increased degrees of genetic or chromosomal damage with aging. There is, in fact, considerable evidence that shows that this occurs. For example, it has been shown that chromosomes from old humans are much more fragile and have a higher number of breakages than those taken from younger individuals. In addition, the amount of tiny nicks within the DNA strand (normally caused by faulty single base-pair substitutions) have also been shown to increase with age. Perhaps the most striking change of all occurs when whole chromosomes are 'lost'. Studies using human white blood cells have shown that the frequency of lost chromosomes increases from 3% in young people (aged 5 to 14) to 9% in persons aged over 65. This may not be a particularly marked change, but it nevertheless shows that genetic damage is a feature of aging.

To provide more support for the theory, however, it needs to be tested experimentally, and one obvious way is to induce mutations in somatic cells and then to see if this speeds up aging. This is relatively easy to do. During the Second World War, scientists discovered that animals subjected to radiation appeared to age very rapidly. Mice, for example, would lose their hair, look debilitated, show decreased activity and die much earlier than untreated animals. Similarly, in humans it is well known that ultraviolet radiation is particularly damaging to the skin, resulting in an aged appearance along with an increased likelihood of skin cancer. Scientists who began to examine the effects of radiation on animals soon found that it produced its effects by damaging DNA. In particular, radiation exposure caused widespread breaks in the chromosomes, along with suspected alterations (or removal) of the bases in the DNA. Furthermore, the more exposure that an animal underwent, the faster its rate of aging, and the more DNA damage it incurred. Thus, radiation caused somatic mutations and clearly provided a most useful experimental tool by which to test the somatic mutation theory of aging.

Unfortunately the somatic mutation theory of aging suffers from one serious experimental disadvantage: while it is easy to measure aging, it is very difficult to observe and to quantify somatic mutations. In fact, it is difficult enough just to observe the chromosomes because for most of the time they are tightly coiled in the cell's nucleus and cannot be detected. The only time they can be observed is when they become uncoiled and separated when the cell divides and replicates itself. Even then, the mutations have to be inferred by counting the number of chromosome breaks or aberrations which may not provide a very accurate measure. On top of all this, many cells of the body, including those of the brain and the muscle cells of the heart, are unable to divide (or replicate themselves) and thus never reveal their chromosomes. Consequently, the somatic mutation theory can never be tested using these types of cell. To examine the effects of radiation on the structure of chromosomes, therefore, most investigators used either skin or liver cells that can easily be made to divide under a microscope. This, of course, is far from ideal when there are many other tissues of the body of greater interest to the aging researcher.

THE WORK OF THOMAS SZILARD

Rather than confront the problem of how to accurately measure somatic mutations following radiation exposure, a number of biologists have attempted instead to develop theoretical and mathematical models in order to explain how somatic mutations can cause aging. One of the first scientists to formulate such a model was the leading nuclear physicist Leo Szilard in 1959. His basic premise was simple. Szilard reasoned that throughout life, the genetic material of our somatic cells were constantly being damaged by "random hits" that either resulted from natural biochemical reactions involving DNA, or from external events such as cosmic rays or ultraviolet radiation striking the bases of the DNA molecule. Thus, in its battle to maintain life, our DNA was gradually and constantly being bombarded and weakened by the chemical and physical events that surrounded it. It is not difficult to picture the potential carnage that such forces could inflict, including the possibility of bases (and base pairs) being knocked out leaving behind empty spaces, or bases being altered or rearranged so that they no longer serve their proper function.

However, this idea is not as straightforward as it first appears. For one thing, Szilard knew that DNA could withstand considerable damage (indeed more than life could reasonably throw at it) and still be fully functional. How did this occur? One probable reason, Szilard believed, was that genes come in pairs (dominant and recessive) and consequently inactivation of one gene would leave the other one intact and functional. Therefore, to completely 'knock out' a gene's function both genes would have to be hit. Szilard's calculations, however, showed that the actual rate of mutation occurring in our cells was too low and the number of cells too large for this to be a realistic theory of genetic decline and aging. Because of this difficulty, Szilard hit upon another solution. He argued that there were certain genes that were more important than others, and the most important genes of all were the ones that controlled the activity and expression of the other genes on the chromosome. Although these 'regulator' genes were relatively few in number, Szilard reasoned that if they were knocked-out they could inactivate whole chromosomes (or chromosomal fragments) thereby effectively putting out of action all its array of genes.

Despite the plausibility of Szilard's theory it has not enjoyed a great deal of experimental support. For one thing, it cannot explain why identical twins have similar life spans. A second objection against the theory has come from experimental studies with animals that have differing number of chromosomes in their cells. For example, the wasp *habrobracon* exists in two forms; those with a single chromosomes in the nuclei of their somatic cells (called haploid), and those with pairs of chromosomes like us (diploid). It is predicted from Szilard's theory that the single chromosome wasps should be much more susceptible to the life shortening effects of radiation because their genes are not arranged in dominant and recessive pairs, and therefore there is an increased probability of a 'lethal hit'. However, the reverse occurs, and it is found that the paired chromosome wasps are the more sensitive to the life shortening effects of radiation exposure.

THE WORK OF HOWARD CURTIS

Although Szilard's work was mainly theoretical, it stimulated extensive research into the link between DNA damage and aging. In particular, it influenced the work of physiologist Howard Curtis who attempted to measure the extent to which somatic mutations accumulated with age. As mentioned above, chromosomes can only be seen (and genetic damage directly observed) when cells are in the process of dividing. Thus, in order to assess genetic damage, Curtis took advantage of the fact that liver cells have the ability to divide when the liver becomes damaged. In short, Curtis destroyed part of the liver in mice of various ages, and then a few days later when cell regeneration was occurring, he removed the liver tissue and examined it microscopically to visualise the dividing chromosomes. The results showed chromosomal damage to be much greater in tissue taken from old animals. In particular, there was an increase in the number of broken chromosomal fragments, along with an accumulation of linked chromosomes that were abnormally joined together. In an extension of this work, Curtis further demonstrated that young mice exposed to ionising radiation also showed an increase in the amount of chromosomal damage similar to that found in older mice. In other words, these findings showed that somatic mutations were a feasible mechanism of aging.

The somatic mutation theory was further supported when Curtis examined the differences in the amount of genetic damage in cells taken from mice that had been artificially bred to be either short or long-lived. For example, in long-lived strains of mice, the number of chromosomal aberrations increased from an incidence of 10% in two month old mice to 35% in 24 month old mice. In contrast, damage in short-lived mice was shown to accumulate much more rapidly with about 20% of chromosomes affected at 2 months of age compared to over 80% at 20 months. Again, this was evidence that somatic mutations accumulated with aging and that they appeared to be linked with the aging process.

However, as with Szilard, not all the evidence supported the somatic mutation theory. For one thing, a number of investigators undertook similar experiments to Curtis, but examined other types of cell and were unable to find any evidence of increased chromosomal damage with aging. Curtis himself also published some research that proved problematical for the theory; in particular, his finding that doses of radiation sufficient to produce a lifetime's worth of somatic mutations in young animals generally did not have any effect on their life span. Moreover, George Sacher showed that low to moderate levels of radiation exposure actually increased the longevity of laboratory animals – a finding that was particularly difficult to equate with the somatic mutation theory. To make matters worse, even when radiation exposure leads to a shortened life span, it is difficult to establish whether this is due to true aging, or because these animals normally have a much higher incidence of malignant tumours. In other words, it is possible that radiation reduces longevity by speeding up certain disease processes rather than accelerating aging.

Although there is little solid evidence for the somatic mutation theory, it has never quite gone away, and some researchers still have a sneaking suspicion that somatic mutations may still contribute to aging (see also the work of Macfarlane Burnet discussed in chapter 10). This uncertainty exists largely because the somatic mutation theory is a plausible theory, yet remains deceptively difficult to prove or disprove. The crux of the problem lies in the fact that somatic mutations are sadly quite invisible even with the most high powered microscopes or today's biotechnology. Consequently, to assess the amount of damage, the scientist is left to count chromosomal breakage, or some other form of gross change, that may not necessarily provide an accurate measurement. Even allowing for this difficulty, however, the mass of evidence is contrary to the theory. Perhaps most damning of all is the fact that the theory cannot adequately explain why identical twins tend to have similar life spans, or why females, on average, live longer than males. Indeed, if aging was a totally random process, as the somatic mutation theory implies, then this would not be expected to happen.

Aging and DNA Repair

DNA is a delicate and highly complex molecule, particularly susceptible to damage from a wide range of different sources. However, it would be wrong to assume that DNA can afford to allow itself to passively accumulate damage as the somatic mutation theory may seem to imply. In fact, quite the reverse is true. In order to protect itself, DNA has evolved a number of remarkable self-repair systems that attempt to correct any damage that is inflicted upon it. These repair systems fall into three main categories: strand-break repair, excision repair and post replication repair; and it is worthwhile briefly describing these protective mechanisms to show just how important the process of repair is to the DNA. The first of these systems (strand-break repair) has the function of mending any breakages that occur in the DNA. This system has to be extremely efficient because it has been estimated that as many as 200,000 DNA breaks occur every *hour* in any given human cell. The second system (excision repair) may be even more important since it acts to snip out any damaged parts (particularly bases) from the DNA, and to replace them with new components. The third mechanism (post-replicative repair) is more specialised and appears to fill in any missing DNA following its replication during the formation of a new cell. All these repair systems are crucial to the integrity of the DNA. Without them our DNA would not be able to exist for any reasonable length of time, and would quickly lose its governing control over the cell's chemical processes resulting in its death.

The importance of DNA repair systems can be illustrated by the disorder *xeroderma pigmentosum* which is due to an inherited genetic defect that prevents DNA repair, particularly in the cells of the skin. This produces a condition in which the sufferer is extraordinarily sensitive to ultraviolet radiation normally present in sun light. People with this disease develop dry scaly skin (xeroderma), severe pigmentation and freckling, and a very strong predisposition to skin cancer. As a result sufferers must remain indoors, and when exposure to sunlight is unavoidable they must carefully cover every part of their body to protect themselves from its damaging effects. Even with these precautions, victims of xeroderma pigmentosum usually die of a skin malignancy before the age of 20 years. There is also evidence that the disorder also extends to other types of cells. In about 50% of cases there is reduced growth and mental retardation, and at post-mortem there are often signs of premature neuronal death in the brain and spinal cord.

The existence of highly efficient and complex repair systems that are continually at work to prevent the breakdown of DNA, gives rise to an alternative version of the somatic mutation theory. Put simply, it has been suggested that the accumulation of genetic damage that occurs with aging may not be so much due to random 'hits', but to the failure of the

DNA repair systems to mend the 'hits'. The consequence again would be the accumulation of genetic damage that randomly builds up in the DNA as the repair systems lose their efficiency, eventually resulting in the breakdown of cells and body tissues as the organism ages.

The first evidence linking DNA repair systems with aging was provided by Ronald Hart and Richard Setlow in 1974, who reasoned that if the DNA repair systems were involved in aging, then there should be a positive relationship between the life span of different types of animal and the extent of their DNA repair capacity. In other words, longer-lived species should have more efficient or higher levels of DNA repair. To examine this possibility, Hart and Setlow took cells called fibroblasts (see chapter 6) from the skin of several mammalian species (shrew, mouse, rat, hamster, cow, elephant and human) that had completed about a twentieth of their life span, and then placed the cells in a nutrient medium that kept them alive and allowed them to divide and replicate. To examine the amount of DNA repair, Hart and Setlow then exposed the cells to various levels of ultraviolet radiation (to inflict DNA damage), and measured the resulting repair by added radioactive thymidine to the cell culture. (Thymidine is one of the building bricks of DNA that is particularly susceptible to damage following ultraviolet radiation, and therefore the amount of new thymidine taken-up and incorporated into the DNA provides a measure of the repair capacity of the DNA). The results of this experiment showed an excellent correlation between thymidine uptake and species longevity. In other words, the longer-lived species had more efficient and active DNA repair systems than the shorter-lived species. Thus it appears that the longer life span of a species may be due to more efficient repair capacity of its DNA. And, perhaps even more fundamentally, it provides further evidence that genetic damage is an important determinant of the aging process (see figure 4.1).

Richard Hart and George Sacher provided further support for this idea when they examined the amount of DNA repair in two strains of mice that had different life spans. The common house mouse (*Mus muscalis*) and the longer-lived deer mouse (*Peromyscus leucopus*) are similar in size, organ weight and gestation time, but differ in average life span by a factor of 2.5 (3.4 years versus 8.2 years respectively). When Hart and Sacher examined the rate of DNA repair in cells taken from these animals they found a ratio of repair that almost exactly paralleled the difference in the life span of these two animals. In other words, the deer mouse had a rate of DNA repair that was 2.5 times greater than its more common cousin.

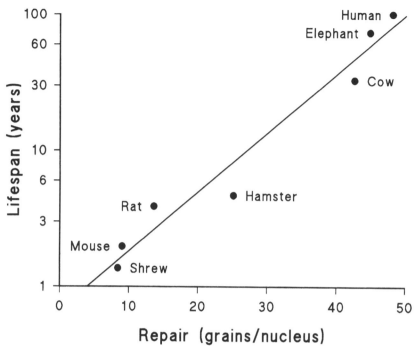

Figure 4.1 The relationship between the extent of DNA repair and life span in 7 mammalian species (data redrawn – not to scale – from Tice, R.R. and Setlow, R.B. (1985) in Finch, C.E. and Schneider, E.L. *Handbook of the Biology of Aging*. Van Nostrand.

These findings raise a further question: could aging across the life span be due to the declining efficiency of DNA to repair itself? In other words, would cells taken from old individuals be less efficient in their ability to correct DNA damage than those taken from younger individuals? This question was examined by Samuel Goldstein who took human skin cells (fibroblasts) from foetal, new-born, young and aged subjects, and exposed them to ultraviolet radiation in order to stimulate DNA repair. The results showed that there was no difference between young and old cells in their ability to repair DNA damage, a finding that has also generally been confirmed with animal research. Do these findings therefore disprove the somatic mutation theory? The answer is no. No matter how efficient DNA repair processes are, there will always be some accumulation of somatic mutations caused by damage that never gets repaired. And, of course, the somatic mutation theory holds that animals who accumulate damage at a faster rate (whether DNA repair declines or not) will be the ones with the shorter life span.

Another way of testing the repair theory is to look at the ability of DNA to repair itself in cells that are taken from individuals with diseases of accelerated aging. As might be expected, cells derived from individuals with xeroderma pigmentosum show practically no repair when they are exposed to ultraviolet radiation. However, cells taken from individuals with progeria and Werner's syndrome frequently, although not always, show normal levels of DNA repair. This strongly suggests that there are other mechanisms underlying diseases of accelerated aging, and some of these will be discussed in chapter 6 which will look at the ability of progeric cells to replicate themselves.

THE REDUNDANCY THEORY

We have seen that only about 5% of our DNA appears to encode for genes and the function of the remaining 95% is largely unknown. Much of this redundant DNA is made up of highly repeated sequences, sometimes containing as many as 200–300 base pairs, that can be repeated 100,000 times or more. However, if redundant DNA can be repeated many times over, then why not the genes themselves? In fact, some genes (although not all) are indeed repeated and there appears to be considerable repetition of genes that code for ribonucleic acid (RNA). The main function of RNA is to copy genetic information from the DNA and translate this message into the manufacture of protein. Indeed, without RNA's assistance, DNA would be as good as useless since it would have no control over the cell (see next chapter). Thus, these two nucleic acids are intimately linked together and this is further shown by the unusual way RNA is formed. Not only does the DNA include the genes that code for RNA, but it also assembles this molecule on its own surface unlike most other genetic products (proteins) that are assembled away from the nucleus in the cell's surrounding cytoplasm.

It appears that RNA is so important to the function of the genome that during evolution DNA has acted to safeguard itself by making back up copies of the RNA genes. For example, it has been shown that the number of repetitions of genes encoding for ribosomal RNA (or rRNA) is around 5–10 in bacteria, 100–130 in insects and between 250–600 in vertebrates. There may also be some redundancy in other genes unconnected to rRNA, perhaps by the backing-up of certain vital codons. These findings have led to an alternative genetic theory of aging first proposed by Zhores Medvedev in 1972. His basic thesis was that the repetition of genes offered considerable protection against random genetic damage; and that the greater the number of genes in reserve, the longer the amount of time they will be able to function before being knocked out of action. In other words, repetition of genes may be a protective device against the accumulation of random errors in DNA.

This theory leads to two predictions: one is that as we age we should gradually use up our quota of repetitive genes; and the second is that we would expect longer-lived species to contain more repetitive genes than shorter-lived species. Both these predictions have been subjected to experimental research. The first hypothesis has been partially confirmed by findings that show a decrease in the number of rRNA genes in certain cells (heart, brain and muscle) taken from aged animals. However, not all cells of the body appear to show this age-related decline, suggesting that rRNA gene loss cannot be used to explain all aspects of the aging process. The second hypothesis predicting increased numbers of RNA genes in long-lived animals has received even less support. For example, Richard Cutler at the University of Texas has looked at a number of species (mice, hamsters, rats, rabbits, chickens and humans) and found no obvious correlation between RNA gene redundancy and aging. In fact, his data showed the rabbit to have the most RNA genes per genome (350) compared to the 220–280 found in the human genome. Clearly, these findings are difficult to reconcile with the idea that gene redundancy is linked to longevity.

It appears that the redundancy theory enjoys little support at present. In fact, it is generally believed that RNA gene redundancy is more likely to be used to increase the production rate of RNA than to defend the organism against aging. In other words, the more RNA that is required by the cell to carry out its functions, the greater the number of RNA genes that will be enlisted in this task. Thus, rather than being redundant, such genes are believed to be functionally active. Nevertheless, the genomes of higher animals contain a lot of 'junk' DNA and it is difficult to believe that it serves no purpose. Although this is purely speculation, it can be argued that it perhaps helps to soak up random genetic damage that would otherwise hit more important parts of the chromosome.

PROGRAM THEORIES OF AGING

Program theories provide a completely different means of explaining how changes in gene function may underlie the aging process. Supporters of this approach believe that aging, instead of being due to genetic wear and tear, is the result of a program that is written into our genes, and which initiates an orderly sequence of body changes as we age. For many researchers this is an attractive idea, particularly as it is not clear how random damage theories can explain many of the basic features of aging; including the fact that each individual species has its own characteristic life span and that significant extensions of longevity can be artificially bred into selected animals. Both of these findings (and there are many other examples that could have been chosen) point to a clear genetic involvement in aging, but not one which is stochastic and operating according to the laws of chance.

Program theories assume that aging is a genetic-developmental process that follows-on from earlier stages of development including growth, puberty and adulthood. Viewed in this way, senescence is simply the final stage of development, and like earlier stages is under its own specific genetic control. Indeed, there is little doubt that much of our development is regulated by genetic factors. For example, in humans, birth occurs after 9 months of gestation, puberty at around 12 years, cessation of growth at about 18 years, and in women the onset of menopause is around 45 years. Furthermore, most of these stages are precisely timed and known to be programed by the switching on and off of specific genes. And, if this is the case then why not old age? Indeed, why should senescence be any different to any of our other life stages? If we accept that it is not any different, then it may be that aging is something that nature has gone out of its way to provide for us, or at least, has sat back and allowed to happen (perhaps by the build-up of deleterious genes later in life – see chapter 2). Either way, if we accept senescence as a genetic-developmental process, then we are accepting the idea that we are programed to decline and age.

If aging is indeed programed in this way then clearly there must be some kind of pacemaker, or biological death clock, somewhere in the body. Some researchers believe that this clock is to be found in every cell of the body, whereas others believe that the timing mechanism is in a single controlling centre such as the brain. These two possibilities are not exclusive, however, since it is also possible that, as Dr Alex Comfort has pointed out, there are several types of clocks which are situated in different levels of the body's organisation. Either way, the program theory is one that tends to capture the imagination largely because, as John Medina points out in his book *The Clock of Ages*, it suggests that a deliberate aging breakdown occurs with banks of "suicide genes" turning on at specific times, causing cells to die and tissues deteriorate. In short, if the theory is correct, our genes turn against us with a vengeance. Despite this, there is remarkably little solid evidence for the theory, although as we shall see below, there are some tantalising clues.

APOPTOSIS

Can it really be that our genes turn against us? At first sight this idea seems improbable. After all, our genes have surely evolved to enable us to reproduce, and to stick around long enough so that we can raise our offspring. Why then should our genes actively turn against us when it might be easier for them to simply fade away and grow old graciously? Indeed, it seems contrary to the evolutionary role of what genes set out to do, for them to be directly responsible for death. Yet recent evidence indicates that at least a few of our genes do indeed have this responsibility. In fact, it appears that every single cell in our bodies contains a built-in genetic program for self-destruction should the need ever arise. This process is called apoptosis (or programed cell death) and rather than being a rare event, it has been shown to be surprisingly common and crucial for the health and integrity of all multicellular life, including humans.

The first realisation that programed cell death was an important feature of life came from studies of embryonic development undertaken in the first half of this century. For example, during the fifth week of human embryonic life our hands look like flippers with no distinct fingers. But then during the sixth week, the webbing of the hand quickly disappears revealing 5 clearly defined digits. What basically happens is that the cells making up the webbing between the fingers simply die, and they do this, not because their blood flow is interrupted, or because they are attacked in some way, but because they are intrinsically programed to do so. Similarly, cell suicide plays an important role in the formation of the brain and spinal cord. At a certain stage of foetal development huge numbers of nerve cells and their projections (axons) are produced. If a particular nerve fibre makes a connection, say with another nerve cell, or muscle, then it will survive. However, on the other hand, if it doesn't (and this occurs is the majority of nerve cells) then the neuron will have its death genes switched-on and it will be killed-off.

Although programed cell death in embryonic development had long been recognised, it wasn't until 1972 when the Australian pathologist John Kerr, and his Scottish colleagues Andrew Wylie and Alastair Currie at the University of Aberdeen, recognised that this form of cell death also occurred in adult organisms; and further, that it continued throughout life. They called the process *apoptosis,* a word derived from the Greek that describes the falling away of petals from a flower or the leaves from a tree. This word is particularly apt as it describes what happens to a cell (as seen through a microscope) when it undergoes programed cell death. An apoptotic cell essentially 'falls away' from all its neighbours, diminishes in size, and disintegrates with its intracellular components forming small dense packets of material (apoptotic bodies) that become ingested by surrounding cells. There is no sign of inflammation and no damage to neighbouring cells. In short, unlike accidental cell death (necrosis) the apoptotic process is very orderly and clean.

It is now known that apoptosis occurs on a huge scale throughout the adult body with millions of cells committing suicide every hour of the day. One place where this occurs is the immune system. The most important cells of the immune system are the various types of white blood cells (or lymphocytes) that circulate around the bloodstream and lymph, looking-out for foreign antigens and invaders. When these white blood cells come across a pathogen they normally multiply in large numbers to help overwhelm the threat. But what happens once the foreign invaders have been destroyed? The excess numbers of lymphocytes are now not needed and it appears that they simply commit suicide by apoptosis. But perhaps even more remarkable is the fact that not only do white blood cells use apoptosis to kill themselves, but they also use the secrets of apoptosis to destroy other body cells. This occurs, for example, when our cells become infected by viruses. These can pose a particularly serious threat to the organism especially if the viruses begin to spread from cell to cell; and to combat this problem the immune system has evolved a special type of cell, called a natural killer cell, whose job it is to recognise infected body cells and to destroy them. One of the ways in which it does this is simplicity itself. In short, it punches a security code into the infected cell that activates its program for apoptosis. In fact, evidence shows that all the cells of our body contain this program for self-inflicted death.

Another place where apoptosis is likely to play a crucially important role is in the detection of cancerous cells. All cells that are capable of division and replication also have the potential for spinning out of control and proliferating endlessly thereby producing tumours. Cancer only develops once a cell has accumulated mutations in several genes that control cell division and growth (see also chapter 6). This is possibly a common event (perhaps even a daily event); but it is generally believed that when such mutations arise, and are irreparable, the affected cell usually commits suicide rather than risk becoming cancerous. Why then does the cancerous cell not follow this escape route? One answer seems to be that the genes controlling the suicide program become damaged. There are probably several genes working in tandem to produce apoptosis (some genes appear to initiate the process whilst others carry it out), but undoubtedly one of the most important is a tumour suppressor gene called p53 that is located on chromosome 17. This gene serves a number of functions including the inhibition of the cell's division, but importantly it is also able to activate the cell's apoptotic machinery when its DNA becomes irreversibly damaged. The importance of this gene (and thus apoptosis) is shown by the fact that many different types of human cancer, including those of the lung, colon and breast, either have damage to this gene, or are manufacturing a useless version of the p53 protein.

How then does the cell go about killing itself? One particularly important class of protein that the cell uses to do this are the proteases. These are enzymes that act to split proteins into smaller components and can be likened to sharp knives which the cell normally takes great care to keep in their sheaths. When apoptosis is set in motion, however, a special set of proteases go to work; and one of the first places they visit is the cells nucleus where they start to slice the DNA into millions of tiny bits. Without its command centre the cell is no longer able to function and its activities quickly come to a halt. However, the proteases continue their work by cutting up the rest of the cell body, whilst other proteins help to bundle up the remains (including the DNA) to form apoptotic bodies that are eventually ingested by neighbouring cells. The cell is therefore dismantled piece by piece.

To what extent is apoptosis involved in the aging process? Nobody really knows for sure, although it is unlikely that apoptosis actively turns against us, and starts to kill our cells in a quest to make us age. It is more likely that apoptotic efficiency declines with age with several knock-on effects including a increased chance of developing cancer. However, the discovery of apoptosis shows that the body does have the potential to turn upon itself should the need ever arise and, of course, if aging is a programed process this is exactly what would be predicted to occur. Apoptosis is not a likely candidate for programed aging (although stranger things have turned out to be true) but if nature can find a way of programing cell death, then why not the programing of aging and the organism's death? In fact, as we shall see below, this is exactly what nature has managed to do, at least with a few of her many species.

Perhaps the best example of where aging and senescence appear to be the result of an intrinsic programed process is in the case of semel-parous animals (and plants) that breed just once before dying. One such example is the silk moth (*Saturniidae*). Many insects including flies and moths do not feed as adults and therefore are destined to quickly die once they emerge from the pupae. However, the silk moth is partic-ularly interesting in this respect because once it has finished laying its eggs, it becomes highly active which appears to serve no purpose other than to speed up its demise. It has been argued that this occurs because the resulting reduced life span makes it more difficult for predators to learn how to recognise the moth, thereby helping to enhance the survival of its offspring. Indeed, supporting this theory is evidence showing that a similar type of silk moth which is brightly coloured (warning predators of its unpleasant taste should it be eaten) does not undergo increased activity once it has laid its eggs and lives longer as a consequence.

However, longer lived species also show semelparous reproduction and one of the best examples is the Pacific Salmon. These fish live in the Pacific Ocean for two or three years before returning to the river where they were born in order to spawn. Within a few weeks of this spawning, however, the salmon literally degenerate and die. They become bloated with a humped back and hooked snout, and the skin becomes festered with large patches of fungi that eats into the scales. There is widespread internal degeneration with most internal body organs showing atrophy and decay. In short, within a space of a fort-night the salmon's physiology effectively runs through a series of events from reproductive fitness to death.

Why then does such degeneration occur? Research undertaken by O.H. Robertson in the late 1950s and early 1960s showed that if the salmon were prevented from spawning then the decay did not occur and resulted in a life span that was often tripled. In addition, Robertson also found that the signal that initiated the deterioration appeared to arise from the testis since if the salmon was castrated, the degenerative process did not occur. Once the chemical signal from the testis was released, however, it acted on another hormonal gland, namely the adrenal cortex, and it was this gland that did all the dam-age. In short, the act of spawning resulted in an outpouring of hor-mones from the adrenal cortex that in turn directly caused tissue damage, as well as shutting down many of the metabolic activities of the body, including the immune system. The evolutionary significance of this programed death is not absolutely clear although it has been argued that the salmon has thrown everything into ensuring the success of reproduction to such an extent that it forfeits the right of surviving afterwards. But even allowing for this explanation it is clear that the process must be programed into the genes in some way. In short, it appears to be a perfect example of programed senescence.

A number of other marine species also pay a similar price for their reproduction including lampreys, squid, cuttlefish and certain types of eel. Another example that has attracted a great deal of attention is the Mediterranean octopus (*Octupus hummelincki*). After laying her eggs the female stops feeding and parts of her digestive tract begin to degenerate. Consequently, she dies a few weeks later of starvation. The key structure in this event is the optic gland which secretes a substance that inhibits the octopus from feeding. If these glands are removed before maturity, not only does the female octopus live two or three time longer, but she grows more slowly as well. Again, the Mediterranean octopus pays a high price for her reproductive responsibilities.

Remarkably, semelparous reproduction has even been found in certain mammals. For example, in Australia there are 6 species of small marsupial mice (*antechinus*) where the males quickly die once they have mated. These mice leave the nest at 8 months of age and establish territories. Three months later the females become sexually receptive and in response to this signal, the males engage in a paroxysm of highly motivated mating and fighting. In fact, as Roger Gosden recounts in his book *Cheating Time*, one animal in captivity was seen to mate with 16 females in succession, two of them twice, before dying from exhaustion! In general, the mating period lasts about two weeks before the mouse dies. As with the Pacific salmon, histological examination shows internal degeneration produced by the secretion of adrenal hormones, and again it has been shown that males who are stopped from breeding, or have their adrenal glands removed beforehand, reap the benefits of a significantly longer life.

However, most mammals, including humans are 'iteroparous' meaning that once we reach maturity we are able to reproduce many times, and at regular intervals, before gradually losing sexual and reproductive vigour. Therefore, it would seem that we do not follow the same pattern of reproductive senescence as semelparous animals although there are grounds for speculating that the situation may not be as clear-cut as it first seems. For example, it is known that older parents produce a higher incidence of genetically abnormal offspring and, therefore, from an evolutionary perspective it might make good sense to shut down the reproductive systems in aged individuals to avoid such calamity. The female menopause may be one such example and evidence suggests that an analogous situation could also occur in some males (i.e., the male 'climacteric'). The female menopause, in particular, would appear to be a good example of a programed aging event in humans, although rather than being switched on by specific genes it occurs because the ovaries simply run out of their allocation of eggs (see chapter 9). Nevertheless, it is conceivable that the changes in body chemistry that results from the cessation of ovulation may in turn be instrumental in altering the expression of other genes, some of which may be involved in aging.

THE EXPRESSION OF AGING GENES

In theory, it might be predicted that the program theory of aging should be relatively easy to test. One strategy would simply be to examine how many genes are 'switched-on' (expressed) through the early part of the life span, and then to see which of these genes become 'switched-off' (or perhaps even 'turned-on') during old age. Unfortunately, this approach is not as easy as it sounds because there are enormous numbers of genes at work in any given tissue; and consequently one becomes faced with a hugely complex situation with the activity of many genes decreasing, some increasing, and others showing no change with aging. Even if genes are found that consistently change their expression as a function of age, this is still no guarantee that they are involved in producing senescence because they may simply be adapting to other age-related changes in the body. In short, trying to identify specific genes and to prove that their expression is inextricably linked with programed aging is fraught with difficulty.

Despite this, molecular biologists have come up with some interesting leads. One of the most intensively studied groups of genes that change their expression with aging are the so-called heat shock genes. These genes produce special types of protein (called heat shock proteins) that protect the cell when it is subjected to unusually high levels of stress. These genes were initially found in 1962 to be activated when the cell was subjected to high temperatures (hence the term 'heat shock') although it is now known that a wide range of stressors including heavy metals, pollutants and various other toxic chemical agents can also induce their formation. These heat shock proteins have a number of important functions in the threatened cell, but in the main they assist in the assembly and insertion of new proteins into the cellular organelles whilst helping out with the removal of their damaged parts. Thus, heat shock proteins provide increased protection to the cell when it is faced with a stressful or emergency situation.

There are several types of heat shock gene and their proteins, but of particular interest is a heat shock protein called HSP70. This protein has been examined in cells taken from laboratory animals, as well as cells kept in culture, and it has been shown that there is a consistent and significant reduction in the formation of this protein in response to stress with aging. Interestingly, other types of heat shock proteins do not seem to be affected. Therefore, for some unknown reason it appears that the gene encoding for HSP70 shuts down (or becomes unable to convey its message) whilst other heat shock genes continue to go about their business. Because it is such a consistent feature of cells taken from aged animals, it is hard to account for this change in terms of random damage and therefore it would appear to be a programed event. This finding has clear implications for aging since it implies that an older organism will be less able to respond to stresses, and as a result will accumulate much more damage in their cells. This, in turn, could turn out to be an important cause of aging at the molecular level.

Does this mean that heat shock proteins are involved in aging? Unfortunately, nobody knows for sure. In fact, the decline in the HSP70 protein could turn out to be an adaptive and beneficial response by the cell in response to some other situation. To give an example: as the cell ages it might consider it prudent to turn-off the HSP70 gene, especially if it is costly to maintain in some way. This might not necessarily turn out to be disadvantageous for the cell because, as we have seen, it also contains other types of heat shock genes and proteins to compensate. This explanation may be rather contrived, but it does illustrate the point that a consistent change in an expression of a gene with aging does not necessarily mean that such genes are involved in aging. Indeed, it is probably the case that aging research has a long way to go before it can confidently link the expression of certain genes (such as HSP70) with the aging process.

Nevertheless, there are some genes where a link with aging seems highly probable. For example, Michael West has helped identify the genes that make the enzymes collagenase and stromelysin, and has found that aging skin cells show increased expression of these genes. Collagenase is an enzyme that helps destroy collagen, and stromelysin is an enzyme that helps break down elastin (both collagen and elastin are important structural components of skin); consequently, the increased expression of these enzymes may be one reason why the skin gets thinner and more flabby with age. One of the exciting implications of this work is that if a means of turning off these genes can be found (perhaps by the development of a special chemical), then it may be possible to keep skin in a permanently youthful condition. Because of these exciting prospects (and the large amounts of money that one day may be made from this research) Michael West has helped form Geron Corporation, the world's first biotechnology company dedicated to developing drugs and therapies to treat aging.

DIFFICULTIES WITH THE PROGRAM THEORY

Despite the plausibility of the program theory there is little solid evidence that senescence is a result of genetic events that are sequentially programed to occur with aging. As discussed above, the difficulties of trying to prove a link between gene activity and aging are very great. However, part of the problem may also lie with the word 'program' itself. A program is generally defined as a set of coded instructions that allows it to perform some operation and, in the case of aging, this implies that the program is wired-in to the genes to produce senescence. But in practice, the word tends to be used much less precisely in aging research. Take, for example, the female menopause, which occurs when the ovaries use up their allocation of eggs. In one sense it can be argued that the menopause is programed to occur at around the age of 45, although this is a rather loose way of describing the situation. Perhaps nature has not really planned on women living past the age of menopause and therefore the event is not programed to occur at all. What looks like a programed event is really a by-product of the design of the human female. The problem is confounded by the fact that cessation of ovulation will change the hormonal status of the body, which will in turn probably change the activity of many genes. Again, the genetic changes that occur will look like they have been programed, whereas in fact they are simply responding to an inadvertent change in the body. Can these changes be regarded as programed aging? In one sense the answer is yes (i.e., the body has been designed that way), but it is not the same type of programed aging that assumes that genes have been designed specifically to produce aging. The first example is in effect a case of inadvertent programing whereas the latter is an example of planned programing. Unfortunately, this distinction seems to be rarely made in aging research (i.e., the word program is not well defined), although it must ultimately be crucial to understanding how aging comes about.

SUMMARY

- Our genome contains approximately 100,000 genes and around 3,000 million base pairs, yet much of our DNA appears to be redundant.
- Genetic theories of aging fall into two main categories: those that assume that genetic decline is the result of random damage, and those that argue that genetic change underlying aging is programed to occur.
- The somatic mutation theory holds that aging is due to the accumulation of damage that occurs to the genes of soma (body) cells. Evidence for this theory has come from experiments showing that radiation exposure (known to cause somatic mutations) reduces longevity and accelerates aging.
- Animals that are long-lived tend to show higher levels of DNA repair than short-lived animals.
- The repetition of genetic information (especially for genes coding for RNA) may provide some protection against somatic mutation or random damage.
- One of the main weaknesses of the somatic mutation theory (and other theories assuming genetic damage) is that it is difficult to explain why identical twins have more similar life spans than non-identical twins.
- Genetic program theories generally see aging as a clock-like process that follows on from earlier stages of development such as growth and puberty.
- Apoptosis shows that cells have a genetic program with the capability to destroy themselves (although this does not necessarily mean that such a process is used to induce aging).
- Some of the best evidence for the program theory comes from semelparous animals which degenerate and die soon after reproduction.
- The expression (or activity) of some genes appears to change with aging. For example, the gene that makes heat shock protein HSP70 is often turned-off in aged cells, whereas the genes responsible for making collagenase and stromelysin become more active in old skin cells (fibroblasts).

THE CELLULAR BASIS OF AGING

The cell, after all, is a machine, so why should it not simply wear out, just as an automobile does?

–Albert Rosenfeld

The human life span simply reflects the level of free radical oxidative damage that accumulates in cells. When enough damage accumulates, cells can't survive properly anymore and they just give up

–Earl R. Stadtman

The last few chapters have provided convincing evidence for the idea that our genetic make-up has an important role to play in the aging process. But genes are just the starting point for producing change in biological systems: and if our genes are to exert any influence, then they must be able to communicate their instructions to a wide variety of other biological molecules in the cell. The molecules that make up the cells of our body (essentially proteins, carbohydrates, fats and nucleic acids) provide a second level of biological organisation by which we can understand aging. Indeed, if we are to fully understand the genetic basis of aging then we must also explain how genes act on these living molecules, and how they set in motion all the chemical reactions that take place in the cell. The most important pathway in this respect is the transformation of a gene into its ultimate end product, the protein molecule. This is a complex process involving numerous chemical interactions, and one that would appear (in theory at least) to be susceptible to increasing decline with aging. Thus, we might expect a decline in the accuracy of protein formation as we age, either because of a change in our genes (see last chapter), or because the actual process of synthesis becomes impaired. However, even if a protein is correctly synthesised, there is yet another possible aging change waiting in the wings. A cell contains a number of internal structures and membranes that are largely derived from protein molecules (and fats) and it might be expected that these cellular structures will be susceptible to wear and tear with continued use. Thus there are two points of attack when we come to examine how cellular function may deteriorate with aging: a decline in the accuracy of protein synthesis, and/or a gradual wearing down of cellular constituents. Both of these processes are likely to result in an impairment of cellular function that in turn may provide an important cause of aging.

THE IMPORTANCE OF PROTEINS

Proteins (the word is derived from the Greek *proteios* meaning 'of primary importance') are vital constituents in all living things from bacteria to man – and without these unique chemicals life as we know it could not exist. Proteins are arguably the most complex and structurally sophisticated molecules known (even bearing in mind DNA), and because of their great complexity they cannot be made from inorganic (non-living) chemical reactions (although certain amino acids can be formed in this way – see later). Consequently, from its earliest origins, life has had to figure out ways of constructing proteins for itself – and this is one of the main reasons why genes evolved in the first place. Genes are simply the blueprints from which proteins are constructed, and if we are to fully understand life and all of its ramifications, we also have to understand how proteins are made and why they are so important.

Proteins are essentially huge and complex molecules consisting of much simpler and smaller subunits called amino acids. To be more precise, proteins are made from long chains of amino acids linked together by peptide bonds that in turn become twisted or folded to form complex shapes. Proteins make up more than 50% of the dry weight of cells and are crucial in almost everything they do. We contain about 60,000 different types of protein each with its own unique chemical structure and specific function. For example, some proteins are used to make the connective tissue and muscle which forms the structural support of the organism. Other proteins are responsible for transporting chemical substances, such as the red blood cells that takes oxygen from the lungs to every cell of the body. Proteins also form hormones (and other messengers) that carry information between glands and various organs. Some proteins can bind to DNA to switch genes on and off, and other proteins make up antibodies that play a crucial role in immunological surveillance and defence. But their most important biological role is, undoubtedly, their role as enzymes that act as catalysts in the chemical reactions upon which life depends. It is a simple truth that life is chemistry, and in this context there is hardly a cellular chemical reaction that does not require the presence of a enzyme made from a protein. The vast majority of the proteins that nature produces are in fact enzymes that are continually at work within our cells every second of our lives.

Proteins come in a great variety of shapes and sizes. One of the simplest is glucagon (a hormone secreted by the pancreas gland), composed of only 29 amino acids linked together in a single chain. In contrast, a complex protein such as collagen (a fibrous protein found in the skin, tendon and bone) may have over 3,000 amino acid links arranged together in several interlocking chains. With so many amino acids having to be incorporated into a protein molecule there are bound to be occasional mistakes. In most instances one or two incorrectly positioned amino acids will not change a protein's overall shape, and it will continue to function normally. But what happens if a protein's shape is altered significantly? A good example is haemoglobin, the protein that carries oxygen in our red blood cells, and consists of 574 amino acids arranged in a chain made up of 141 links. There only needs to be one substitution of a single amino acid in the chain at *exactly the right point,* and the whole shape of the protein molecule becomes changed to such an extent that it can no longer effectively pick up and transport oxygen around the body. In fact, this is precisely what happens in the genetic disease sickle-cell anaemia. Similarly, cystic fibrosis is another disease where the inclusion of just one faulty amino acid is enough to alter the basic shape of a single protein with devastating consequences.

Sickle-cell anaemia and cystic fibrosis are extreme examples but, nevertheless, they show how a simple change in a protein's structure can have catastrophic consequences for the individual. Of course, most of us do not suffer from inherited genetic disease, and we begin life with a normal set of cells and their constituent proteins. However, as we saw in the previous chapter, our genetic material is not foolproof, and is constantly subjected in the process of living to damage and somatic mutations. Furthermore, the conversion of a gene into a protein provides another way in which protein modification can occur. Both these processes could, in theory, result in the random incorporation of faulty amino acids that gradually changes the shape of our proteins over time and alters their functional properties. And it follows that, if enough cells were affected in this way, then the organs and body systems of which they are part might begin to show functional decline and aging. Therefore, it is possible that proteins have a potentially vital role to play in the biological decline of the organism. The important question is do they?

How are Proteins Made?

The coded instructions for making proteins are to be found locked away as genes in the cell's nucleus, but the sites of protein manufacture (as we shall see) are to be found located outside in the cell's cytoplasm. Thus, information from the gene must somehow flow from the nucleus to the cytoplasm. But how does this occur? In short, the answer lies with another type of nucleic acid called ribonucleic acid or RNA. There are several types of RNA, but the one with responsibility for transporting the DNA's instructions into the cell where protein formation takes place is called, appropriately enough, messenger RNA (mRNA). In short, the process of protein formation can be described simply in what is known as the central dogma of molecular biology: DNA is *transcribed* into mRNA which in turn is *translated* into protein. Thus, mRNA acts as the crucial intermediary between the DNA and protein manufacture. Without this messenger, DNA would be locked away in the nucleus and helpless to control events in the cell.

Although RNA is a close cousin to DNA in terms of its chemistry (its sugar is ribose instead of deoxyribose, and the base uracil is substituted for thymine), the most important difference between the two molecules is that RNA is much smaller, and exists only as a single strand (unlike DNA which is doubled-stranded). This enables RNA to move freely in and out of the nucleus, as well as providing a template by which to copy transcripts of the much larger DNA molecule. How does genetic information encoded in the DNA get transcripted into RNA? The process begins with a special enzyme, called RNA polymerase, which attempts to bind to a part of a gene known as the promoter region (if the promoter region is blocked by a repressor protein then gene transcription will not occur). If RNA polymerase is successful in binding to the promoter, it will unzip part of the DNA molecule exposing a few of its bases, and in the process start to make a strand of mRNA. It does this by travelling along the exposed bases, grabbing a complimentary base from the watery nuclear surroundings, and linking it to the backbone of the ribose strand using similar (but not identical) base-pairing rules as for DNA replication (see chapter 2). The process is not unlike the threading of beads on to a string to make an open-ended necklace. When RNA polymerase comes to the end of the gene it recognises a sequence of bases that tells it to stop its activity, at which point it literally falls off the DNA and begins its new function as a cellular messenger.

The mRNA then leaves the nucleus coded with its mirror image copy of the gene and enters the cytoplasm where it seeks out a special structure called a ribosome, which is the site where the assembly of the protein will take place. Ribosomes are sometimes described as 'work benches', although they actually resemble large ball-like structures made up of protein and another type of RNA called ribosomal RNA (or rRNA).

It appears that the rRNA helps position the mRNA and then, once it is pinned-down, it exposes the mRNA's bases (called codons) three at a time. As the bases on the mRNA strand become exposed they immediately become hooked-up with amino acids. There are only 20 natural amino acids found inside the cell and these are hunted down by yet another type of RNA called transfer RNA (tRNA). As well as carrying amino acids to the ribosome, the tRNA molecule also contains a specific three-base codon that is able to pair with the codons being exposed one at a time on the mRNA. As a result, the tRNA picks up the appropriate amino acid and transports it to the ribosome where the two codons are zipped together, thus joining the amino acid to the mRNA. Enzymes within the ribosome then create a peptide bond (to make a link for the next amino acid); following which the next mRNA codon is exposed, and so on. This process continues until all the bases in the mRNA have been filled up, at which point a protein is made.

The construction of a protein is an incredibly efficient process. It has been estimated that ribosomes bond approximately 10 amino acids per second to a strand of mRNA, and that it takes around a minute to make an average protein. Furthermore, this process is highly accurate with ribosomes only making one error in around 10,000 amino acids, or put another way, one single mistake for every 2 dozen proteins! Considering that our cells are continually synthesising proteins, and that more than 1 million peptide bonds are made *every second* in an

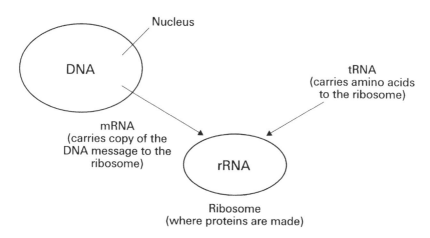

(DNA is transcribed into mRNA which is then translated into protein)

Figure 5.1 A simple summary showing the steps involved in protein formation.

average mammalian cell, this degree of efficiency is astonishing. This high rate of synthesis is needed because the human body contains tens of thousands of different proteins that are continually being broken down and replaced every second of the day. Many proteins only exist for a few minutes, being built when needed and destroyed soon after. Other proteins may last several hours or days, and in a few cases (such as the crystalline protein found in the lens of the eye) they may even last the lifetime of the individual. The way in which proteins are broken down is just as important, although the process is not fully understood. One mechanism appears to involve a protein called ubiquitin, which binds to unwanted or discarded proteins, and in the process attracts a class of marauding and destructive enzymes called proteosomes that begin to dismantle the doomed protein into its constituent parts. Life is a continual flow of protein formation, and in most cases as one protein is broken down another one will be taking its place (see figure 5.1).

THE ERROR CATASTROPHE THEORY

As we have seen, from start to finish, protein synthesis involves a number of complex chemical steps, and it would perhaps be surprising if some age related decrement did not occur in this process. For example, errors could occur in the copying (transcription) of the genetic information from the DNA to the RNA; in the creation of the various types of RNA; or in the assembly (translation) of the protein at the ribosome. Indeed, any error along the DNA-protein pathway could result in an incorrectly made protein that might impair its function and have adverse consequences for the cell. It hardly takes a great deal of insight to imagine that these types of error could be accumulating with aging, particularly as somatic mutations in the DNA (see chapter 4) might be adding a further source of error to the process.

There are, however, several problems with this apparently simple theory. As already mentioned, what is critical to a protein's function is not so much its sequence of amino acids, but rather its shape – which is created by the twisting and turning of the amino acid chains. And in most cases an incorrect amino acid inserted into a chain will have little impact on the protein's shape and function. Even if the mistake does result in the protein being pushed out of shape, any faulty proteins tend to be quickly broken down and replaced by new ones. Thus, random or chance errors occurring in the formation of proteins, even if they increase with age, should not pose too much of a problem for the cell. In other words, aging might be expected to decrease the efficiency of protein synthesis, but not necessarily alter the final working state of the proteins themselves.

However, in 1963, a biochemist called Leslie Orgel proposed an alternative theory to explain how a build-up of faulty proteins might occur with aging. Orgel pointed out that certain proteins were much more important in the running of the cell than others, and the most important of all were the enzymes involved in the transcription and translation of DNA into protein (one such enzyme is RNA polymerase although there are others). An error in this type of enzyme would be very serious since it would lead to a continual flow of incorrect information from the DNA to the ribosome, resulting in proteins that were riddled with mistakes. Orgel called such an event an "error catastrophe" because once protein synthesis was contaminated in this way, then the errors would quickly multiply overwhelming the cell with faulty proteins. The situation is not unlike having a conveyor-belt full of faulty proteins that cannot be turned-off, and which floods the cells with its damaged products before they can be cleared away. The end result, Orgel reasoned, would be the death of the cell; and if enough cells died in this way, then the net result would be a decline of physiological function recognisable as aging.

PROBLEMS WITH THE ERROR CATASTROPHE THEORY

At first sight it might appear that the main prediction of the error catastrophe theory should be relatively easy to test. In short, it predicts that there should be a build-up of faulty proteins in cells taken from aged individuals, with such changes resulting in aging. Unfortunately, despite the fact that this theory has generated an enormous amount of research in the field of biochemistry over the last 30 years or so, these predictions have remained extremely difficult to prove. The main problem lies with the proteins themselves. Because of their extremely small size, and complex three-dimensional shape, it is not possible to visualise a protein in the same way as say a set of lego bricks, even with an electron microscope. Because of this, biochemists have to infer the changes in a protein's structure by measuring its biological activity in laboratory tests. There are a number of tests although most involve purification of an enzyme that is then used to catalyse (or speed up) a specific and measurable chemical reaction (these are highly technical experiments that are practically unfathomable to anybody except a biochemist). However, put simply, if the catalytic properties of an enzyme show a decline with aging, then this is an indication that the shape of the enzyme molecule has been altered in some way.

As Melvin Green points out in his book *Classic Experiments in Modern Biology*, the purification and analysis of proteins probably consumes more laboratory hours per person than any other endeavour in modern biology. In short, it is a long-winded and laborious process. Despite this, thousands of such experiments, involving hundreds of different enzymes taken from many tissues of species of various ages, have been undertaken to examine whether the efficiency of enzyme activity declines with age. In general, very little conclusive evidence of such a decline has been found. In fact, it appears that about 80% of the enzymes so far examined show no decline in activity with aging, while the remaining 20% show both increases and decreases in function. Indeed, most researchers now agree that, in general, proteins taken from old cells perform with the same fidelity as proteins from young cells. This is contrary to the error catastrophe theory which predicts that, by the end of the life span, there should be a significant build-up of all types of abnormal protein.

Another line of evidence against the error catastrophe theory has come from experiments in which analogues of amino acids, or RNA bases, have been fed to laboratory animals. Analogues are substances that are sufficiently similar to normal amino acids, or bases, to enable them to be incorporated into new proteins, or RNA molecules; but are different enough to impair the function of the molecules that contains them. Consequently, not only should the incorporation of these analogues into protein, or RNA, lead to an accumulation of faulty proteins, but if the error catastrophe theory is correct, it should also result in a significantly shortened life span. Despite this, there is little evidence to suggest that the life span of experimental animals are affected by analogue rich diets in this way.

It is fair to say that the majority of investigators do not believe in the error catastrophe theory as a mechanism of aging. Nevertheless, there is a small degree of doubt. For example, Robin Holliday has shown that DNA polymerase enzymes (which are involved in the replication of DNA) obtained from human cells, appear to show age-related alterations that are in accordance with the error catastrophe theory. Furthermore, there is some support for the accumulation of abnormal proteins with aging in bacteria and yeast. Despite this, most biochemists accept that there is much more evidence against the theory than what supports it. But perhaps the real crux of the matter lies in the fact that scientists do not have the means to measure very accurately what happens to proteins with aging. For this reason, a definitive test of the error catastrophe theory remains elusive.

Although the accuracy of protein formation does not appear to decline with age, the rate at which proteins are synthesised does. In general, the overall rate of protein formation declines by about 40–70% over the life span in mammals, and an even greater decrease is found in insects and other organisms. In other words, the older cell is able to make proteins correctly, but it takes longer to do so. In addition, aged cells appear to be much less efficient at removing old or defective proteins and, for this reason, proteins in older cells tend to have a longer life than those found in young cells. This finding poses another difficulty for the error catastrophe theory because if proteins are existing for longer in the aged cell before being removed, then it is much easier to explain the occurrence of faulty proteins (should they be found) in terms of damage (i.e., wear and tear) that takes place *after* the protein has left the ribosome. Again, this shows the great difficulty in trying to prove error catastrophe as a plausible theory of aging.

THE FREE RADICAL THEORY OF AGING

An alternative approach to explaining how cells might accumulate damage, both to their proteins and other cellular constituents, is the free radical theory of aging. This theory (which has already been mentioned in previous chapters) was first proposed by Denham Harman in 1956 who argued that aging was the result of intracellular damage that came from chemical reactions involving free radicals. A free radical is any chemical with an odd number of electrons in its outer shell, although in biological systems these chemicals nearly always derive from reactions involving oxygen. An oxygen atom normally has six (or rather three paired electrons) in its outer shell which makes it relatively stable; but should the oxygen molecule lose any of these electrons in a chemical reaction, then it becomes highly unstable. An atom containing a free electron does 'not like' being unpaired and consequently it will move frantically around the cell until it finds another molecule to latch on to. Although the lifetime of a free radical is measured in thousandths of a second, they are nevertheless capable of doing considerable damage during this time. In effect, they can knock bits out of proteins, lipids and nucleic acids, as well as causing molecules such as DNA to cross-link (see below) where they could be an important cause of genetic damage.

Most free radicals derive from reactions that occur in the course of the cell's metabolism involving oxygen. Oxygen is essential to the energy-producing reactions that keep us alive, and its metabolism is more or less a constant process in the tissues of the body. Although most oxidative reactions take place within special isolated structures of the cell known as mitochondria, free radical products can nevertheless leak from these structures where they set up harmful chain reactions throughout the cell. In fact, mitochondria are not exempt from oxidative damage, and this has also been suggested as an important cause of aging (see chapter 7). It is perhaps ironic to think that the element we depend on so much for life has such a deleterious effect. But this is the price we pay for generating the energy to keep us alive, and most forms of life are stuck with this arrangement. To make matters worse, the cell can also be subjected to free radical damage from a number of other sources including radiation and toxic chemicals. Thus the cell is under constant free radical bombardment from every direction. In fact, our cells produce literally millions of free radical molecules each and every day. Bruce Ames and his colleagues at the University of California have estimated that the number of oxidative hits to DNA per cell per day is about 10,000 in the human, with at least 10 times this number of hits occurring to the mitochondria. Miraculously, our cells somehow withstand this onslaught, helped in part by the protection of antioxidant enzymes (see below).

All chemical constituents of a cell are susceptible to free radical damage, but a group of substances known as polyunsaturated fatty acids are especially vulnerable. Free radical attack induces peroxidation of unsaturated fatty acids (a similar process occurs when butter goes rancid), and this in turn tends to set-up chain reactions of free radical formation that cause considerable damage. Polyunsaturated fatty acids have a crucial role to play in the cell because they are a vital component of its various membranes. Most structures in the cell are surrounded by special membranes including the nucleus, mitochondria and lysosomes (the latter being small sacs of digestive enzymes that normally function to dissolve foreign particles entering the cell or to break down cellular components that are no longer needed). Thus, free radical damage to cellular membranes could impair the transport of information in and out of the nucleus, render the mitochondria incapable of producing energy, or cause the lysosomes to become leaky so that they spill their damaging digestive enzymes into the cytoplasm. The cell itself is also surrounded by a highly specialised membrane, and consequently the transport of substances in and out of the cell itself might also become impaired as a result of free radical damage.

In short, the free radical theory of aging proposes that the structural and functional changes we recognise as aging are a result of intracellular damage (including damage to DNA, membranes and proteins) produced by free radical reactions. In addition, free radicals have also been suspected of causing and aggravating a number of degenerative diseases of aging including cancer, cardiovascular disease, immune-system decline, arthritis and Parkinson's disease (in fact more than 60 disorders have been implicated as free radical diseases). Although aging and disease are separate processes, the two are nevertheless inextricably linked, and clearly one advantage of the free radical theory is that it has the potential to explain both.

DEFENCES AGAINST FREE RADICAL ATTACK

Fortunately, our cells are not helpless in the fight against free radicals, and they have developed powerful protective enzymes to combat their damaging effects. These enzymes probably began to evolve with the development of life around 3.5 billion years ago in order to protect cells against the high levels of ultraviolet radiation that existed then; and again more recently (some 1.3 billion years ago) to protect life when concentrations of oxygen began to rise in the Earth's primeval atmosphere. Thus, it is probably the case that free radical protection evolved hand in hand with life itself. Indeed, without such protection it is hard to imagine how life could have existed for any length of time.

There are a number of important intracellular enzymes that fight free radical reactions, and some of these, including superoxide dismutase, catalase and glutathione have been implicated in causing aging and mentioned in previous chapters. Each of these enzymes is involved in mopping-up and combating a specific free radical reaction, and they often work in tandem to help each other out. For example, superoxide dismutase converts the relatively common free radical called superoxide into oxygen gas, along with a second free radical called hydrogen peroxide which is in turn broken down by catalase into oxygen and water. Our cells also contain a number of patrolling antioxidants that we obtain from our diet, including vitamin E (alpha-tocopherol) and vitamin C (ascorbic acid), which also have a similar function. Vitamin E is a particularly important as it is fat soluble and inserts itself into the cellular membranes where (as we have seen above) free radicals have the potential to do considerable damage. Vitamin C is water soluble and an effective free radical scavenger in the watery compartments of the cell including its cytoplasm.

If free radicals cause aging, and if life has had to rely on the protection of antioxidant enzymes to exist, then this raises the interesting question of whether the variations in life span that exist between different species might be linked to the intracellular levels of these enzymes and antioxidants. This possibility was examined in a study by

Richard Cutler and his research student Julie Tolmasoff in 1980 at the National Institute of Aging. They measured the levels of superoxide dismutase taken from the liver, brain and heart tissue of ten different types of primate (ranging in age from 12 years in the tree shrew to over 95 in the human), as well as two types of mouse. Surprisingly, they found no relationship between cellular levels of superoxide dismutase and life span. For example, whilst levels of superoxide dismutase were very high in man, they were also high in the much shorter-lived marmoset and lemur. Moreover, although superoxide dismutase levels were very low in the mouse, they were almost as low in the gorilla which lived about 15 times as long. Thus, on first sight the results did not appear to make any sense; if longevity did indeed evolve then it seemed it did so without the help of superoxide dismutase.

However, upon further consideration, these results are not as strange as they first appear. The important difference between the species chosen by Cutler is not so much their life span, but rather their metabolic rate which is a measure of how quickly a tissue uses energy and therefore oxygen (see chapter 7). A mouse, for example, uses up energy far more quickly than a gorilla. Therefore, if both have similar levels of superoxide dismutase it should mean that the gorilla will actually live longer because it will produce fewer free radicals per unit of time. In other words, the superoxide dismutase will provide much more protection in the gorilla than a mouse. Thus, one would not necessarily expect a simple relationship between life span and superoxide dismutase activity. However, one would predict a relationship between levels of superoxide dismutase and total energy expenditure over the animal's entire lifetime. In short, the best combination for predicting increased longevity (such as man) is a slow metabolism (or oxygen use) accompanied by high superoxide dismutase protection, and this is indeed what Cutler found when he recalculated the data to test his hypothesis (see figure 5.2).

These findings lend considerable support to the idea that evolution has had to balance the level of metabolism against antioxidant protection to overcome the life shortening effects of free radical damage. Indeed, as we saw in chapter 3, genetic mutations in the nematode worm that either increase the level of antioxidant enzymes or lower the rate of metabolism, have also been found to increase life span, thus lending further support to Cutler's findings. In other words, the evolutionary evidence strongly suggests that free radicals are an important cause of aging, and this principle would also appear to apply on an individual level. Therefore, aging may occur because our antioxidant defences are not perfect; or because the efficiency of these enzymes declines with time as with many other systems of the body. Either way

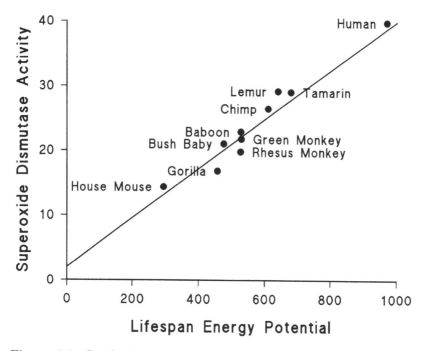

Figure 5.2 Graph showing the longevity relationship between levels of superoxide dismutase and life time energy expenditure (data redrawn – not to scale – from Cutler, R.G. (1983) *Gerontology*, 29, 113–120.)

we would predict an accumulation of free radical damage with aging, which eventually produces the physiological decline we recognise as senescence.

THE EFFECTS OF ANTI-OXIDANTS ON THE AGING PROCESS

If the free radical theory is correct, then it should be possible to increase the life span of animals by feeding them antioxidants that inhibit or slow down free radical reactions. In fact, this is relatively easy to do because as we have already seen, both vitamin E and C are powerful cellular antioxidants and there are many other chemicals that have similar properties. Indeed, a large number of experimental studies have now looked at the effects of antioxidant administration on the life span of a wide range of animals as diverse as plankton, worms, fruit flies and rodents. This work was initiated in 1957 by Denham Harman who tested several antioxidants, and found that one in particular, called

2-mercaptoethylamine (2-MEA), increased their average life of mice span by 30% if added to their diet. This increase was, according to Harman equivalent to raising the average human life span from 73 to 95 years.

However, a consistent finding from Harman's work, as well as others, is that whilst antioxidant administration often increases the *average* life span of treated animals, it does not increase their *maximum* life span. In other words, antioxidants allow more animals to reach the end stage of their natural life span but not beyond, suggesting that the treatment has protected the animal from early mortality. Indeed, if 'early deaths' are omitted from the mortality figures of untreated groups of animals; or alternatively, if great care is taken to enable non-treated animals to reach their maximum life span, then groups of antioxidant treated animals are rarely shown to live longer than controls. These findings suggest that antioxidants may be providing protection against disease rather than aging. Indeed, there is some evidence to support this view since a number of studies have found that animals treated with antioxidants develop significantly fewer tumours. Other research has indicated that antioxidants may also be having a beneficial effect on the immune system since they slow down the process of auto-immune disease in genetically susceptible animals. Thus, it appears that antioxidants do not necessarily slow down aging, but rather they provide protection against a number of age-related diseases.

Nevertheless, even this work needs to be viewed with some degree of caution. For example, Harman has only managed to extend average life span in mice by feeding them with what Alex Comfort has described as "heroic doses" (that is, up to 1% of the food eaten). In addition, there may be other confounding variables at work helping to extend the animal's life span. For example, in 1971, Alex Comfort examined the effects of the antioxidant ethoxyquin on the longevity of mice, and found that this drug increased the life span of mice by 18% in males and 20% in females. However, during the course of this study, Comfort also went to the trouble of regularly weighing his mice and found that they showed a significant weight loss throughout their life. This is a problematical finding because it is known that food restriction can also increase longevity (see chapter 11). Indeed, other researchers including Harman himself, have on occasion reported decreases in food consumption in antioxidant treated animals. One is confronted by the possibility, therefore, that the administration of antioxidants to experimental animals, especially in the high doses used by many researchers, may make their food less palatable, or even reduce food intake by causing nausea.

HUMAN STUDIES

Of course, it is not possible to perform similar rigorously controlled studies, involving huge doses of antioxidants over the entire life span of human subjects. But there is accumulating medical evidence that diets high in antioxidants can have beneficial health effects. A particularly interesting study in this respect was published by James Enstrom and the Nobel Prize laureate Linus Pauling in 1982. These researchers followed, over a 6 year period, a group of 479 elderly Californian men and women (all of whom were over the age of 65 years at the start of the study) who had replied to a questionnaire placed in the health magazine *Prevention*, and admitted to regularly consuming high amounts of vitamins and mineral supplements. At the beginning of the study in 1974, the average daily intake of vitamin E for the group was 500 international units (IU), and the average intake of vitamin C was 1.7 grams; both of which had also been typically consumed in high quantities for several years prior to the study. The results showed that the people who consumed vitamin E had a 50% lower death rate, and a significantly reduced level of heart disease compared to non users. Despite these impressive findings, however, the results also showed that their was no advantage to be obtained by increasing the daily amount of vitamin E intake over 100 IU, and that very high doses of vitamin E (greater than 1000 IU) appeared to be actually detrimental to extending life span. Unfortunately, the results from this study examining the effects of vitamin C consumption were inconclusive – largely due a poor return of questionnaire data.

We must be cautious in interpreting the results of this study since its participants were almost certainly not representative of the general population, and not only differed in terms of vitamin intake, but also in a number of other respects. For example, readers of *Prevention* magazine were probably more likely to have eaten balanced and healthy diets; to be more knowledgeable about health matters; and to be motivated to keep fit. And clearly, all of these factors, regardless of vitamin consumption, could have also contributed to their increased life span. In short, we can not say with any certainty whether the vitamin intake had a beneficial effect on longevity in this group of people.

Nevertheless, evidence continues to mount showing the beneficial effects of antioxidant administration on health. Perhaps the most impressive research of all supporting this position was published in the prestigious *New England Journal of Medicine* in 1993, which reported on two studies examining the benefits of taking vitamin E as a protective measure against heart disease. Both studies followed large groups of people over a long period of time (4–8 years) with subjects regularly filling-in questionnaires that recorded dietary practices and patterns of disease. One study led by Eric Rimm and his colleagues gave questionnaires to 39,910 men aged 40 to 75 years, and the other study led by Meir Stampfer gave a similar questionnaire to 87,245 women aged 34 to 59 years. What was most important, however, was the fact that both groups of subjects were free of heart disease at the beginning of the study. Any patterns of heart disease that developed in these groups could therefore be examined in light of dietary practices. The most striking finding to emerge from this work was that large doses of vitamin E (above 100 IU) was associated with a significantly decreased risk of heart disease in both groups. Overall, subjects who regularly took vitamin E showed a 30–40% less incidence of heart disease compared to age-matched control subjects. In contrast, high intake of vitamin C, was not found to have any beneficial effect on heart disease.

The beneficial health effects of vitamin E has also been demonstrated in a British study, published in 1996, by Nigel Stephens and his colleagues at the University of Cambridge. These researchers examined over 2,000 subjects with heart disease who were recommended to take vitamin E following their diagnosis. The results showed that vitamin E (taken in amounts of 400–800 IU daily) reduced non-fatal heart attacks by 36% and fatal heart attacks by 66% when used for between 1 and 2 years. Although the doses of vitamin E were very high, and there is still considerable uncertainty over the most effective dose to be used, these results provide highly convincing evidence that vitamin E has a protective effect in heart disease.

If we are to summarise all the evidence, both animal and human, concerning the effects of antioxidants on longevity, then it must be that they appear to offer protection against disease but do not increase maximum life span or indeed slow down aging as the free radical theory predicts. In fact, considering the great weight of evidence that supports a free radical involvement in aging, then such a conclusion is perhaps surprising. One can only imagine either that it is not biologically feasible to boost cellular antioxidant levels to such an extent that they offer complete protection against free radicals, or that there are other mechanisms of aging at work that are independent of free radical damage.

Despite this, there appears to be relatively little harm in taking high amounts of antioxidant vitamins, and perhaps not surprisingly, several popular and well written books, including *Life Extension* by Duke Pearson and Sandy Shaw, and *Stop Ageing Now!* by Jean Carper, advocate vitamin and mineral consumption as a means of benefiting health and longevity. But perhaps even more importantly, a number of prominent scientists engaged in aging research also appear to agree. For example, the Nobel laureate Linus Pauling who died in 1994 aged 93 years, used to regularly take 18,000 milligrams of vitamin C each day (this is equivalent to about 250 oranges!), and believed that such a regime could add an extra 12 to 18 years to one's life. Similarly, Roy Walford who has undertaken pioneering research on the immunological basis of aging, and the effects of calorie restriction (see chapters 10 and 11), includes among his daily regime, 1,000 milligrams of vitamin C and 300 IU of vitamin E. But perhaps we should leave the final word with Denham Harman who pioneered the free radical theory back in 1954. Among the vitamins he consumes each day are 2,000 mg of vitamin C (taken in four 500 mg doses); 150 to 300 IU of vitamin E; and regular daily administration of beta carotene, ubiquinol (coenzyme-10), selenium, zinc and magnesium. One can also not be failed to be impressed by the fact that at the time of writing, Harman is 78 years old and still remains emeritus professor of medicine at the University of Nebraska – apparently a living testimony to his free radical theory.

THE WASTE ACCUMULATION THEORY

A cell is essentially a chemical factory in which simple substances are built into more complex ones (anabolism), and complex molecules are in turn broken down to make simpler ones (catabolism). But, like any chemical factory, there must be waste products that occur as a result of these processes, and the constant production of free radicals might also be expected to add further wreckage to this waste. Thus, the cell needs to have an efficient means of waste disposal. But what happens if such a process is less than perfect? This brings us to the waste accumulation theory of aging which proposes that senescence is the result of intracellular debris that builds up over time, and which eventually reaches a level where it impairs the ability of the cell to fulfil its duties.

The most visible waste product to be found scattered throughout the cytoplasm of aged cells is a yellowish-brown granular pigment called lipofuscin. This substance was observed as long ago as 1842, and was quickly regarded as a 'clinker of metabolism' that accumulated with time. Lipofuscin can be found in practically all cells of the body although it is to be found in its highest concentrations in long-lived cells that are unable to divide or replicate, such as heart muscle and brain cells (these types of cell are known as 'post-mitotic' and will be discussed in the next chapter). Lipofuscin has been shown to accompany aging in every species examined and, furthermore, its rate of accumulation appears to match the rate of aging of that organism. For example, the rate of lipofuscin deposition in the dog heart has been found to be approximately 5 times faster than in the human heart; a difference that roughly corresponds to the life span of these two species. Similarly, most strains of nematode worm live up to 28 days, yet their cells contain approximately the same amount of lipofuscin as those taken from very old humans.

The origin of lipofuscin in the cell is still far from clear, although it is widely believed that it may arise as a result of free radical damage of various cellular membranes that contain a high amount of lipid (fat). Alternatively, it could be that lipofuscin results from the degeneration of mitochondria, or even be the left-over insoluble residues of lysosome activity (lysosomes are enzymes that digest a wide range of cellular material). In short, the origin of lipofuscin is not known. Chemically, lipofuscin is a complex substance that typically contains a lipid core surrounded by various proteins and carbohydrates that have been tightly cross-linked (see below) or fused together. The strongest evidence for a free radical involvement in the formation of lipofuscin has come from studies in which laboratory animals have been fed longterm diets that are deficient in vitamin E. The consequence of this dietary manipulation is that deprived animals typically show a much higher rate of lipofuscin accumulation in their cells compared to controls. Since vitamin E is known to protect the cell against oxidative damage, it seems reasonable to conclude that lipofuscin is a by-product of free radical damage. However, not all evidence supports this theory. For example, one might equally expect that animals fed diets rich in vitamin E should show a reduction in the accumulation of lipofuscin, although this is not the case.

A number of studies have demonstrated a clear relationship between lipofuscin and aging. For example, in 1957, Bernard Strehler found that the amount of lipofuscin occupying heart muscle cells in humans increased linearly with age, so that by the age of 90 years, up to 7% of the cell's interior (or cytoplasm) was occupied with this pigment. Even more striking is the accumulation of lipofuscin in the nerve cells of the brain. Although there is considerable variation between different brain areas, it has been found that some brain cells (notably those in an area called the cerebellum) are almost completely full of lipofuscin in people who are over 100 years old. There is even one area of the brain (the inferior olive) where it has been reported that 85% of the cells are completely full with lipofuscin by the age of 38 years. In fact, the extent of lipofuscin accumulation is so marked in these cells that it results in the nucleus being pushed away from the centre of the cell to its periphery.

It would be reasonable to assume that the clogging up of the cell with lipofuscin is likely to interfere with its metabolism, or function, and thus be a cause of aging. Surprisingly, however, most of the available evidence does not support this view. For example, brain neurons taken from old individuals that are full (or nearly full) of lipofuscin show normal function when tested with electrophysiological recording techniques. Since the main function of a neuron is to send electrical messages, this is strong evidence that its role in communication has not been compromised by the lipofuscin. Similarly, the amount of lipofuscin in heart tissue does not appear to be related in any way to heart disease. Thus, lipofuscin may simply be a relatively harmless by-product of living, and something that the cell 'can work around' in much the same way as people can work around a cluttered apartment. However, there is some degree of uncertainty over the issue since there is a rare human genetic disease called Batten-Vogt Syndrome whose main feature is the marked accumulation of lipofuscin in the brain, spinal cord and optic nerve. Victims of this disease rarely live beyond 20 years, and suffer from mental deterioration and blindness.

There have been some attempts to reduce the build up of lipofuscin by using various experimental drug therapies but, unfortunately, these have turned out to be very disappointing. This is perhaps not surprising since lipofuscin has been shown to be highly resistant to degradation. For example, extracts of lipofuscin taken from experimental animals have been shown to be resistant to laboratory procedures using powerful solvents such as alcohol, ether and chloroform. It is hardly surprising, therefore, that our own cells are unable to break down this aging pigment. Thus, once lipofuscin has been formed, we appear to be stuck with it as an inevitable consequence of aging.

THE CROSS-LINKAGE THEORY

Cross-links are simply chemical bonds that occur between large complex molecules that lie in close proximity with each other. In some cases they may form naturally when two molecules come together and intertwine but, in most instances, they arise as a result of interaction with other chemical agents that react and fuse together parts of the adjacent molecules. Whatever the cause, cross-linkages alter the physical and chemical properties of the molecules they link, rendering them less flexible, and therefore less efficient at fulfilling their biological roles.

Again, this theory overlaps to some degree with the free radical theory because oxidative reactions are known to produce cross-links, and since these links are known to accumulate with aging. But the cross-linkage theory is much broader in scope than the free radical theory for two reasons. Firstly, all sorts of chemical reactions (not just those involving free radicals) can induce cross-links and secondly, the build up of cross-links not only takes place in the cell, but also importantly, can take place *outside* cells (a possibility that is often overlooked in aging theories). Thus, the cross-linkage theory is much more than a variant of the free radical theory.

The cross-linkage theory was first proposed by the Finish industrial research chemist Johan Bjorksten in 1963 who noticed the similarity between the "darkening" that occurred as a gelatine photographic film ages (gelatine is in fact a protein derived from the boiling of animal's hides and bones), and some of the changes that occurred with aging in living organisms (e.g., the thickening and darkening of skin). Bjorksten knew that the tanning of photographic film was due to the formation of cross-links that "stiffened" the molecules of gelatine; and he also knew that a similar hardening process occurred with many other substances over time including leather, rubber, paints, plastics and paper. Indeed, the phenomenon of cross-linking was no secret for inorganic chemists. But, nobody had realised that such a process could also apply to biological tissue, that is, until Bjorksten saw the feasibility of the idea.

Since our cells are basically containers full of large and complex proteins that have long flexible strands enabling them to fold into complex shapes; and since there is a constant flow of chemical reactions taking place in the cell, one can readily see that the cell provides a perfect medium for cross-links to occur. Indeed, as we have already seen, one of the reasons why the cell is not able to break down lipofuscin is because its constituents have been so tightly cross-linked together that it can not be dissolved by the digestive juices of the lysosomes. But it is likely that all biological molecules, to some extent, are susceptible to the damaging effects of cross-linking. To give an example: if two DNA molecules line up side by side (e.g., when the cell is about to divide), they form a perfect target for any molecule that happens to drift along and is capable of reacting with both DNA strands. This type of cross-link is potentially fatal for the cell since it could result in it being unable to replicate or, even if successful, cause DNA breakage and mutation.

To fully appreciate the cross-linkage theory, however, one needs to realise that cells do not make up all the tissue in the body. About a quarter of the human body is composed of connective tissue whose main function is to bond together all of its various cells and structures. Some parts of the body such as bone and skin are comprised almost totally of connective tissue, whereas in other structures this tissue provides the framework for holding the cells (such as those in the liver and heart) in place. One of the most important components of connective tissue is a substance called collagen which provides a major constituent of skin, blood vessels, bone, cartilage and tendons. Collagen is not a living cell as such, but a long thin protein molecule (consisting of three coiled strands wrapped around each other like a rope) that makes up about 30% of the total protein content of the body. It is secreted in thin strands by special cells called fibroblasts (see chapter 6) that are found embedded in the cellular matrix of various body structures. Other types of connective tissue, including elastin and reticular fibres, also provide further support as well as having more specialised roles. For example, the effective pumping of the heart requires that the arteries expand and rebound from the beat in order to add extra force to the blood flow, and this elasticity is largely due to the elastin fibres found in the vascular wall.

One of the most important characteristics of collagen fibres is their ability to hook themselves together to form longer and thicker strands. In general, these links add strength to the collagen containing tissue and may give it some degree of flexibility and elasticity. But unfortunately, with aging, collagen continues to accumulate cross-links which significantly reduces its pliable properties. This can be observed most clearly with the skin. If the skin on the back of the hand is pinched between thumb and forefinger and held for 5 seconds, once released it will quickly revert back to its normal state in a young person but take much longer in an old person. However, there are more ominous aspects of collagen cross-linking than that which occurs with the skin. For example, the cross-linking of collagen may be responsible for the reduction in oxygen capacity of the lungs, or the reduced elasticity of the artery walls which makes them more rigid and susceptible to rupture. Reduced elasticity will also impair the circulation and flow of the blood; while the capillaries that contain collagen will tend to become less permeable, thereby slowing the transport of gases, nutrients, hormones and toxic waste products across their walls. At the cellular level, increased rigidity of cellular membranes will also impair the functioning of the cell, and cross-linking in the cell itself may well be an important cause of DNA damage. In fact, viewed in its widest perspective, cross-linkage could also conceivably be seen as the cause of somatic mutations, and as the main cause of the faulty translation of proteins leading to error catastrophe.

Not only do the number of cross-links increase with aging, but the actual *amount* of connective tissue making up the tissues also increases with age. Collagen is continually produced throughout the life of the organism, and it appears that when specific cells and tissues are injured and die they are often eventually replaced with collagen or other types connective tissue. Thus, the accumulation of connective tissue in many organs of the body (along with a decrease in the number of cells) provides yet another indicator of aging decline.

EVALUATION OF THE THEORY

What evidence is there to show that an increase in the cross-linking of molecules occurs with aging? Unfortunately, it is not possible to see cross-links under a microscope so the biologist has to resort to other techniques to infer their existence, and measure their properties. One of the simplest methods is to see how much newly formed collagen can be extracted from animals of various ages. Newly formed collagen does not have much time to create cross-links and can be readily extracted from tissue by soaking it in salt solutions. Indeed, the results show that as the animal gets older the amount of readily soluble collagen decreases indicating that it is becoming increasingly cross-linked. Another way of examining the properties of collagen is to measure how easily dissolvable it is by using certain enzymes known as collagenases that exist in the body to help break down connective tissue. The ease with which this process occurs is another good measure of the rigidity and degree of cross-linkage and, as might be expected, aged tissue including that which has been derived from human accident victims is much more resistant to this type of degradation.

What are the main factors that cause cross-links to occur? The answer is almost anything! As long as there is enough collagen, adequate temperature for chemical reactions to occur, and sufficient time, then cross-links will form. However, there are a large number of chemicals found in the body that significantly speeds up their formation including free radicals, aldehydes (these occur in minute amounts but nevertheless are very potent cross-linkers), citric acid and sulphur. Another important agent that induces the formation of cross-links is glucose and this may be one reason why young diabetics (who have high levels of glucose in their blood) often show many changes associated with advanced aging, including hardened and clogged arteries, reduced kidney function and stiff joints. In fact, glucose is also known to react directly with proteins and DNA to form a class of intracellular waste known as advanced glycation end products that, in turn, provide a further source of potent cross-linking agents. These substances are increasingly being seen as important determinants of cellular decline and aging.

According to Bjorksten, the cross-linkage theory is the most successful theory of aging yet developed. Although few gerontologists today hold the cross-linkage theory in such high esteem, it nevertheless remains an important theory, and one that is nicely compatible with other explanations of aging including somatic mutation, error catastrophe, free radical and waste accumulation theories. However, one of its weaknesses is that it does not easily explain why different species have different life spans. For example, the collagen taken from a three year old rat is much more densely linked than a collagen taken from an three year old child. In one sense this is no surprise, but if cross-linking was a simple time-dependent process then this should not happen. Clearly, therefore, other factors such as diet, hormone level and rate of metabolism must play an important role in determining the speed of the cross-linkage process. And if this is so, then this suggests that cross-linkage rather than being the cause of aging, is instead *secondary* to other aging changes taking place in the body. Such is the difficulty of trying to pin down the causes of aging!

Nevertheless, if Bjorksten is correct and aging is directly (or indirectly) due to the formation of cross-links, then anything that can slow down, or even reverse, their formation might have benefits for aging. One approach has been to develop drugs with the ability to inhibit cross-links. Sadly, although a few compounds of this type have been produced, none have had any appreciable effect on increasing the life span of experimental animals. In fact, the only thing that appears to slow down the formation of cross-links is food restriction. Not only do animals placed on calorie restricted diets consistently live longer, but they also appear to have much more 'youthful' collagen compared to animals of the same age that have unlimited food. This is particularly strong evidence for the cross-linkage theory and shows that diet is an especially important determinant of how quickly or slowly cross-links can be formed. This issue is discussed further in chapter 11.

SUMMARY

- Proteins are complex molecules consisting of one or more chains of amino acids that are folded into complex shapes (it is the shape of the protein that determines its function).

- DNA is transcribed into mRNA, which in turn is translated into protein at the ribosome. The formation of a protein also requires the assistance of rRNA and tRNA.
- The error catastrophe theory predicts that there will be a build up of faulty proteins with aging – particularly if errors occur in enzymes that are crucial for protein formation.
- In 1956, Denham Harman proposed that aging was the result of intracellular damage due to the formation of free radicals. It has been estimated that each cell's DNA receives around 10,000 free radical 'hits' each day, and that this figure is probably 10 times higher in our mitochondria where most free radicals are produced.
- Our cells are protected by a number of antioxidant enzymes including superoxide dismutase, catalase and glutathione. In addition, vitamins C and E also act as powerful antioxidants.
- In general, it appears that groups of laboratory animals given diets rich in antioxidants show an increase in average longevity (possibly due to a decline in disease) but no extension of maximum life span.
- Large scale human studies have provided strong evidence that vitamin E can provide protection against heart disease in both men and women.
- Lipofuscin is a waste product that accumulates in cells (particularly those in the brain and heart) with aging. Whether it is linked to deterioration of function is not known.
- Cross-links are chemical bonds that join together molecules (including proteins), and which have a tendency to increase in number with aging. They can cause intracellular damage as well as reducing the flexibility of connective tissue.
- About a quarter of the human body is composed of connective tissue whose main function is to bond together all of its various cells and structures. One of its main constituents is the protein collagen (an important component of skin, blood vessels, bone and cartilage) that also has a strong tendency to cross-link.

THE HAYFLICK LIMIT

The greatest desire of man is for eternal youth

–Alexis Carrel

*Dogmas did not die in the last century. They are still present.
Many of today's cherished beliefs will surely fall before some
young iconoclast tomorrow.*

–**Leonard Hayflick**

All living creatures from the simplest to the most complex are made-up
of cells. The human body, for example, is composed of approximately
ten million million cells, of many different types, all of which go together
to make-up its various organs and tissues. Bearing this in mind we
might ask: what is the typical life span of a cell? This apparently simple question – one which is crucial to an understanding of the aging
process – is, in fact, deceptively difficult to answer. In the course of this
book we have seen that DNA damage, error catastrophe, free radical
reactions, waste accumulation and cross-linkage can all impair the
cell's function. And, of course, the obvious end point of all these damaging processes must be the death of the cell. Thus, the cell is mortal and
one may expect this to correlate in some way with the aging of the
organism. However, there is a catch: most cells in the body (although
not all) are able to replicate themselves, which means that they are
able to make new copies of themselves before they wear out. Viewed in
this light, cells therefore appear to have the potential to be 'immortal',
because if they get old, they simply replace themselves. Perhaps this
capability should not surprise us because, as we saw in chapter 2, the
cells that carry our genetic inheritance (i.e., our germ cells) have to be
'immortal' in this way to allow the continuation of the species to take
place. But what about the rest of our bodily (somatic) cells? Clearly, in
an evolutionary sense, it is not so crucial for these cells to be renewed
ad infinitum because they are genetic dead ends and not required to
pass their genes on. Nevertheless, this makes it all the more intriguing
to know whether our body cells are destined at some point to stop
dividing; or whether they are able to replicate themselves throughout

life indefinitely. As we shall see, the answer to the question has an important bearing on how we understand the aging process.

ALEXIS CARREL AND THE INVENTION OF TISSUE CULTURE

To show that a cell has the potential to be mortal (or immortal), it is first necessary to find a way of keeping it alive outside the body. The first person to show that this was possible was Ross Harrison working at John Hopkins University in 1907. Harrison was interested in nerve tissue growth and in order to examine this process he attempted to keep a fragment of tadpole spinal cord alive on a glass slide by immersing it in its own juices. Using this method, Harrison found that he could keep the tissue alive for several days allowing him to detail (with the aid of a microscope) the various stages of nerve growth. Even more important, however, were the general implications of this discovery for Biology. It was soon found that several types of tissue could be kept alive 'on glass' providing the culture medium contained the right type of serum and balance of nutrients. Within a few years the tissue culture technique had become commonplace in laboratories all around the world, and a wide variety of tissue including kidney, bone marrow, spleen and thyroid found a new *in vitro* existence in glass dishes and flasks. Moreover, not only were cells capable of being kept alive, but many types of cell were also able to replicate themselves, enabling vast numbers of new cells to be created and closely examined.

The person who pioneered the tissue culture technique more than any other was the French physician Alexis Carrel, who won the Nobel Prize in 1912 for his work on the surgical transplantation of blood vessels. One of the first issues that arose with this new technique concerned the means by which cultured cells were being kept alive. Some investigators believed that cultured cells were utilising the nutrients in their medium, whereas others thought that the cells were simply using up their reserves of food already stored within themselves. One way of tackling this problem was to grow cultures over long periods of time and this is what Carrel set out to do. Using cells taken from the heart tissue of embryonic chicks, he soon succeeded in keeping the tissue alive indefinitely. This did not mean, however, that on an individual basis his cultured cells lived for ever. Individually, these cells were probably mortal in the long term, but whilst alive they endlessly replicated and made numerous identical copies of themselves. It is in this respect that the tissue in the laboratory flasks appeared to be immortal since it was self-propagating and never died out.

Keeping cells alive in this manner, however, proved no easy task. Because the cells were constantly replicating, they needed to be carefully tended and regularly sub-cultured (sub-divided) so that their number could be kept within manageable limits. This required considerable work; over a 10 year period Carrel reported that his cells had been sub-cultivated some 1,860 times. Furthermore, Carrel kept his cells in a culture that regularly required cleansing with special solutions to remove the build up of waste products, following which it was necessary to reconstitute the medium using fresh chick plasma. The task of looking after these cells became a full-time occupation, and it also evolved into a strange, almost religious, ritual, with Carrel's technicians wearing full-length flowing black hooded robes every time the cells were tended. Using these somewhat unconventional laboratory practices, Carrel's set of cultured cells were kept alive for 34 years before they were finally discarded and allowed to die (two years after Carrel's death). During this time, an astronomical number of cells had been created. In fact, J.A. Witkowski has estimated that had every one of the cells produced by Carrel been allowed to proliferate without limit, then the volume of accumulated cells over the 34 years would have been greater than the size of the solar system! (Witkowski 1980, page 134).

The conclusion that clearly followed from Carrel's 34 years of painstaking research was that our soma cells are capable of continual replication. This had important implications for understanding aging because it indicated that the causes of aging did not spring from *within* the cell. Carrel had apparently shown that providing the culture medium was carefully tended, and renewed at regular intervals, then the cells would carry on replicating for ever. This suggested that cells did not age by themselves, and consequently the cause of cellular aging must ultimately lie *outside* the cell. Researchers speculated that perhaps the extracellular fluid in which the cells were bathed got contaminated in some way, or perhaps, there was a hormonal or even immunological cause to the aging of cells. Whatever the reason, Carrel's research suggested that his soma cells were like their germ cell cousins: in other words, given ideal conditions they were capable of continuous division and replication.

Despite this, a few cell types were found to be incapable of dividing once they reached their fully developed (or 'differentiated') state. For example, brain cells (neurons), heart muscle cells, and the lens cells of the eye were not able to replicate themselves, and it is for this reason they are called post-mitotic (meaning 'past-division'). Indeed, if such cells are lost during the course of life they can not be replaced (this is one of the reasons we tend to lose brain cells with aging). However, most cells of the body are 'mitotic' meaning that they are capable of division and renewal, and typically these are the types of cell (especially taken from embryonic tissue) that are easy to grow in the laboratory. Therefore, Carrel's theory that cells are intrinsically immortal applied to most, although not all, of the somatic cells of the body.

THE WORK OF LEONARD HAYFLICK

The work of Alexis Carrel held great influence; few doubted his findings or the conclusions that logically derived from them. However, this all changed in 1961 with the publication of a now classic paper by Leonard Hayflick and Paul Moorhead. During the 1950s, Hayflick was working at the Wistar Institute in Philadelphia, examining whether viruses could induce cancer in healthy human cells. His strategy was simply to cultivate human tissue taken from aborted foetuses, and then try to induce cancer cells by exposing them to certain viruses. During the course of the work, Hayflick found that his cultured cells grew and divided for many months, but then, to his great surprise, they began to slow down and eventually stop, following which they soon died. Hayflick replicated his work several times with similar results. He knew on account of Carrel's work that this should not occur, but his results were clear cut: that is, human cells appeared to have a finite life span and could not replicate themselves continually as Carrel had shown.

Hayflick attempted to publish his findings in the respected *Journal of Experimental Medicine* but the paper was rejected on the grounds that cultured cells, kept under the right conditions, were known to be immortal! Such a reply, of course, implied that Carrel's work was beyond doubt, and that Hayflick's method was in some way flawed (the assumption being that his cells had been kept under less than perfect conditions). Hayflick's paper was then submitted to *Experimental Cell Research* where it was accepted for publication. In fact, Hayflick's findings have now been replicated many times over by numerous other researchers, and there is now little doubt that cultured cells do have a *finite* life span. In other words, cultured cells are not immortal after all – they stop dividing after a number of divisions, and eventually die out.

Nobody knows for sure what caused Carrel's mistaken results, but Hayflick believes that the problem lay with the culture medium in which the cells were kept. Carrel used plasma taken from chick embryos to culture his cells, and according to Hayflick it is probable that this was

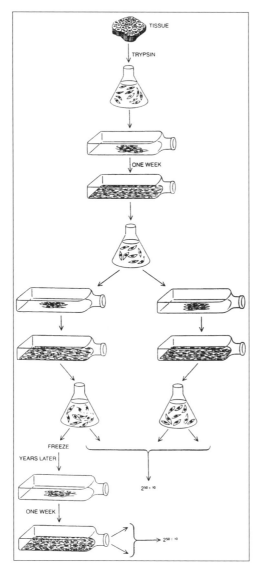

Figure 6.1 Diagram showing the technique used by Hayflick to culture fibroblast cells. Firstly, human cells are broken down into individual cells using the enzyme trypsin. They are then transferred to a flat bottle and allowed to multiply until they cover its bottom surface – at which point they stop dividing. The cell population is then divided into two equal halves which are then re-cultured. This process can only be repeated about 50 times with human foetal cells, and less with cells taken from older individuals. Illustrated by Alan D. Iselin, *Scientific American* 1990. Reprinted with permission.

contaminated with a few young isolated cells derived from other parts of the body that began to replicate when placed in the culture medium. Whatever the reason, this example illustrates vividly how scientific truths can be tied to the prevailing fashions created by figures of authority, that in turn lead to entrenched ideas and dogma. This is not the first time this type of situation had occurred in the history of science and it almost certainly won't be the last.

The technique used by Hayflick to culture cells is relatively straightforward. First, small pieces of embryonic lung tissue are removed from a foetus and exposed to a digestive enzyme (trypsin) which separates the tissue into millions of individual cells. The cells are then separated from the trypsin, and placed into bottles containing a nutrient solution which are then incubated at body temperature for a few days. This causes certain cells in the tissue called fibroblasts to divide and replicate (fibroblasts are cells found throughout the body, particularly the skin, that secrete the protein collagen and other types of connective tissue). After a few days the cells begin to spread out across the floor of the bottle, and after about one week the cells reach a state of confluence where they cover the entire bottom surface of the bottle one cell thick. When the cells reach the wall of the vessel they stop dividing, a phenomenon known as 'contact inhibition'. At this point the cells are removed from the culture, broken down again using trypsin, and distributed in equal measure into two new bottles containing fresh nutrient medium, and the whole process starts again. Although difficult to visualise, Hayflick has shown that statistically, each cultivation represents on average, one cell doubling itself (see figure 6.1).

THE HAYFLICK LIMIT

The crux of Hayflick and Moorhead's crucial discovery in 1961 was that fibroblast cells could only replicate a finite number of times before they stopped and the cell lineage died out. But even more important was the finding that the number of replications a fibroblast was able to undergo before dying was highly predictable. On average, Hayflick and Moorhead found that their cells were only capable of dividing about 50 times, over a period of 8–12 months, before they stopped completely. There was some degree of variability since some cells only survived 40 doublings whereas others occasionally managed to reach a total of 60 divisions; but in general the average was 50 and the cells never went beyond the range of 50 ± 10 divisions. This figure is now widely known as the Hayflick Limit, and shows, in contrast to the long held traditional view supported by Carrel, that cells are mortal and have a very predictable and finite capacity to replicate themselves.

Closer examination of the cultured fibroblast cells by Hayflick showed that they passed through three main stages during their life span. The first phase represents the initial growth of the cells in their original flask where they adapt to the new culture. Following this, the cells enter the second phase, lasting some 6 months or more, where they show vigorous cell proliferation. Finally they enter a terminal phase where they reach the end of their life and ability to replicate. This change, however, does not occur suddenly, but rather the cells begin to take longer and longer to divide as they approach the limit of 50 doublings; and it is at this point that Hayflick believes that the process of aging is setting in. Finally, the cells stop dividing completely regardless of how long the culture is incubated. By this stage most of the cells have become noticeably larger and they begin to show a number of degenerative changes, although a small number of cells continue to survive for some time in this terminal state. These three stages are not only confined to fibroblasts since all normal dividing cells of the body appear to have their own Hayflick limit including cells taken from skin, muscle, liver and kidney.

DETERMINANTS OF LIFE SPAN

The discovery of the Hayflick limit raised the possibility that aging was the result of intrinsic changes occurring within individual cells and, furthermore, that the fibroblast method was an important experimental means by which to examine this process in more detail. Consequently, Hayflick's discovery was the starting point for a large number of studies to test this new model of aging. One of the first questions that Hayflick looked at was whether the limited life span of fibroblast cells was a result of the number of replications they had undergone, or simply due to the passage of chronological time. To answer this, Hayflick suspended his cells in subzero temperatures by storing them in liquid nitrogen after a certain number of divisions. The results showed that when these cells were unfrozen they exhibited a remarkable memory. For example, if the cells had been frozen at their 20th population doubling, and later thawed, the cells would continue to divide about 30 more times and then stop. If they were frozen at the 10th division, the cells in contrast underwent about 40 more replications before stopping. In other words, the total number of divisions always conformed to the Hayflick limit (50 ± 10); that is, the amount of time in storage has no bearing on the cell's divisional capability. In fact, one of Hayflick's human cell strains, called WI-38, retained perfect memory after more than 24 years of preservation in liquid nitrogen.

Another way of examining this phenomenon is to place cells under conditions where no division occurs (by changing the content of the culture medium) but where maintenance is adequate for keeping the cells alive. Again, it appears that the critical determinant of how long a cell lives is not chronological time, but rather the number of divisions that it has undergone. Thus, fibroblast cells do not measure the passage of time, but rather they contain some kind of counting mechanism which records the number of replications. Of course, under natural conditions, this counting mechanism probably provides a reliable estimation of time, and it is only when the cell is experimentally manipulated in the laboratory that evidence of the internal counting mechanism is exposed.

THE EFFECTS OF DONOR AGE

If the Hayflick limit is truly relevant to human aging then it must follow that fibroblasts taken from old adults should not be able to divide as many times as those taken from young or embryonic tissue. This is a crucial test, and to examine this, Hayflick compared lung fibroblast cells taken from human embryo cells with those taken from adult post-mortem donors varying in age from 26 to 87 years. The findings showed that the adult fibroblasts appeared to produce fewer replications, although there was considerable variability in the results. Hayflick showed that, on average, the adult cells divided about 20 times (the results ranged from 14 to 29) in comparison to the average of 50 for the embryonic tissue. Despite this, Hayflick was unable to show a precise correlation between the age of the donor and the number of cell divisions. Other investigators, however, have managed to show this relationship in a much more convincing way. For example, George Martin at the University of Washington in Seattle, examined skin fibroblast cells taken from a hundred subjects of varying ages and showed that there was a decrease of 0.2 divisions for each year of the donors age between the ages of 10 and 90, although once again, there was considerable variability in the results. Thus, as a general rule, it does appear that as the individual ages the capacity of their cells to divide declines.

A particularly interesting variation of this type of experiment has been undertaken on individuals who are victims of the premature aging diseases of Progeria and Werner's syndrome (see chapter 3). If these disorders are genuine diseases of premature aging, then we might expect that cells taken from afflicted individuals should show less capacity to divide than those taken from normal individuals. In general, the results have confirmed this prediction but, again, there is a disconcerting amount of variance in the data. For example, in a study where skin fibroblasts were taken from a 9 year old progeric child, researchers were only able induce two replications before they died. In contrast, some studies using progeric cells have managed to produce between 30 and 40 replications, which although less than usual, is not as severe as might be expected. A more consistent pattern of results has been reported using cells taken from subjects with Werner's syndrome. Most studies have looked at cells taken from individuals who are between 40 and 60 years of age, and nearly all have shown a consistent reduction of around 2–12 cell divisions compared to normal subjects. Although the difference may not appear to be marked it is nevertheless significantly less than normal.

DIFFERENT SPECIES

One of the most fascinating ways of applying the Hayflick limit to aging research is to see whether it tells us anything about the differences in life span that exist between species. If the Hayflick limit is relevant to aging then we might expect longer-lived species to have the ability to produce more cell divisions than short-lived animals. One of the earliest studies to test this possibility was undertaken by Samuel Goldstein who examined fibroblasts taken from Galapagos tortoises (believed to be the longest lived species on Earth). Goldstein took cells from 4 tortoises (2 old and 2 young) and found that they reached their Hayflick limit after between 90–130 doublings. In contrast, a mouse with a maximum life span of 4 years produced a maximum of only 28 doublings. When combined with human data, these results seemed to support the general rule that the longer-lived the animal, the greater its Hayflick limit. A number of other species have since been examined, and although there are some marked discrepancies, the results tend to show that long-lived species do indeed generally produce more cell divisions. For example, Dan Rohme at the University of Lund in Sweden has tested skin cells taken from a number of animals including rats, minks, rabbits and horses, and found a clear and direct relationship between the number of cell replications and life span of each species (see figure 6.2).

Another way of testing this relationship is to examine cells from closely related animals that differ markedly in life span. As mentioned

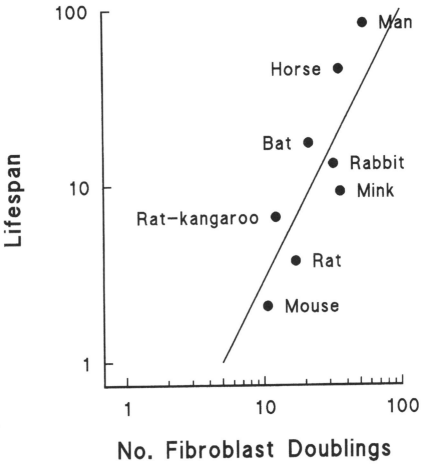

Figure 6.2 Graph showing the relationship between the number of fibro-blast doublings (Hayflick limit) and length of life span for 8 mammalian species (data redrawn – not to scale – from Rohme, D. (1981) *Proceedings of the National Academy of Sciences*, 78, 5009–5013.)

in chapter 4, this approach has been used by George Sacher and Ronald Hart to compare the level of DNA repair in two different strains of mice, *mus musculus* and *peromyscus leucopus*, that differ in average longevity by a factor of 2.4 (3.4 years versus 8.2 years respectively). Not only did Sacher and Hart show a difference between these two strains of mice in terms of DNA repair but, importantly, they also demonstrated that fibroblast cell cultures taken from *promyscus* produced 2.5 times more cell divisions than the same cells taken from the *mus* mice. In other words, there was a direct relationship between the life span of the two mice and the total number of divisions that their fibroblast cells could undergo in culture.

WHERE IS THE CLOCK LOCATED?

As Hayflick and others have shown, there is a natural limit to the cell's capacity to divide and reproduce, and this limit seems to be fixed by the cell itself. If this is the case, then it is hard to escape the conclusion that there must be a program somewhere in the cell that is counting the number of divisions that it has made, and is therefore directly controlling the aging of the cell (and, of course, perhaps the aging of the organism). The question therefore arises as to the whereabouts of this counting mechanism. Hayflick attempted to address this problem by developing a technique that enabled both cellular cytoplasm and nuclei to be separated, and then transferred and reinstated into different cells. The logic behind the technique was quite simple. If the program was located in the cytoplasm, then cytoplasm taken from old cells and infused into new cells should limit their ability to replicate. Alternatively, if the counting mechanism was situated in the nucleus, then the infusion of old cytoplasm into the new cell would have little bearing on the number of divisions that it could make.

This type of experiment is technically difficult to undertake and was only made possible by the discovery of a drug called cytochalasin in 1967, that caused the cell's nuclei (normally found in the centre of the cell), to move to the cell's edge and to extrude through the cellular membrane. In 1972 Hayflick showed that if cytochalisin treated cells were subjected to high **g** forces in a centrifuge, then the nuclei could be removed from the cell completely, leaving behind cultures of enucleated cells (called cytoplasts). These cells remain alive for several days, during which time their contents can be fused into new cells containing nuclei.

Hayflick found that young cytoplasts fused into older cells with nuclei had relatively little effect on their ability to divide although old senescent cytoplasts did produce a small diminished growth potential in young cells. Overall, however, the results suggested that the most important determinant of cellular division was located in the nucleus, and presumably therefore the genetic material. Hayflick examined this idea further by extracting nuclei taken from cells of various ages and transferring them into new fibroblasts. According to Hayflick the results were similar to the cytoplast studies, and showed that old nuclei tended to reduce the replicative capability of their new hosts. Unfortunately the interpretation of these experiments have been difficult because it is impossible to transfer nuclei without small amounts of cytoplasm also being transferred. To summarise: the results strongly indicate that it is the nucleus which contains the allimportant program, although a cytoplasmic influence cannot be ruled out completely.

TISSUE TRANSPLANTATION

By its very nature, the measurement of the Hayflick limit can only be undertaken *in vitro* (i.e., outside the living animal). Critics of this procedure have pointed out, quite rightly, that what happens in a glass flask may not be the same as what happens in a living animal (*in vivo*). The two situations are clearly very different. Unfortunately, it is impossible to measure the Hayflick limit *in vivo* because there is no way of keeping track of individual cells. Not only are our cells so incredibly small, but it is difficult to imagine how they could be individually tagged and identified, let alone a means devised that would measure the number of times such cells had divided. But, there is one way that the problem can be tackled, and that is by transplanting live tissue from one animal to another, and then to see how many times this procedure can be repeated. Although this technique does not measure the Hayflick limit as such, if the Hayflick limit is at work then there should be a limit to how many times this type of procedure can be carried out.

One of the first researchers to examine the viability of transplanting tissue into new animals was Peter Krohn at the University of Birmingham who, in 1962, began transferring sections of mouse skin to new recipients using hair colour as means of identification. The results showed that skin grafts could be successfully applied for several generations, although they could not go on for ever. For example, from a total of 305 initial transplants that were transferred over 3 generations, 32 were still identifiable at 7 years, 7 survived for more than 8 years, and one piece of skin lasted for over 10 years. Although these skin grafts outlived their original donors by some years, the results nevertheless indicated that cells *in vivo*, like their *in vitro* counterparts, had a finite life span.

However, on closer inspection the results are not as conclusive as they first appear. One of the problems that occurred during Krohn's experiment was that the skin tissue got progressively smaller as it was transplanted from animal to animal, so much so that it eventually got too small to transplant. This was partly due to the encroachment of scar tissue; although it also appeared that the remaining 'good' tissue, to some extent, was also getting smaller with each transplant. If this was indeed the case then there were two possible explanations. Either the transplanted tissue was not proliferating normally because its cells were coming to the end of their Hayflick limit, thus making the tissue smaller; or the new *in situ* conditions were not optimal for skin growth. This latter explanation was a real possibility since most cells *in vivo* need to be connected to blood vessels in order to obtain oxygen and nutrients, and many types of tissue need some degree of innervation by nerve cells. The extent to which this occurred with Krohn's skin grafts was difficult to judge, but nevertheless the implication was that, had the capillary and neural innervation been better, then the cells could have lived longer.

Krohn also undertook experiments that examined the effects of transferring old skin to young recipient mice, and vice versa. If the Hayflick limit is a reliable indicator of aging, then it follows that old skin (which presumably has undergone more divisions) should survive for less time than young skin when transferred to new mice. However, Krohn found that there was little difference between the ability of old and young donor skin to grow in their new hosts, a finding that appeared to directly contradict Hayflick. But again, other explanations are possible. For example, it has been pointed out that the old skin grafts may have been replenished by cells migrating from their young host which, of course, is somewhat similar to that which may have occurred with Carrel's cell cultures several decades before. In short, transplantation research involving skin is fraught with methodological problems.

To circumvent some of these problems, Krohn also examined the effects of transplanting ovaries into female host mice. In many mammals (although not humans) the ovaries fail to function at a certain age despite the fact that there are plenty of eggs (or oocytes) left. The consequence of this failure is that the oestrous cycle closes down, and the female no longer becomes periodically 'on heat' or capable of becoming pregnant. The obvious question to ask, is therefore: what causes this decline? Is it the eggs that have aged in the ovary, or is it the chemical environment of the body that has changed? To answer this question, Krohn looked at the outcome of transferring old ovaries into young hosts, and vice versa. The results of this study showed that the old ovaries implanted into young mice reinstated the oestrus cycle, but when the ovaries were taken from young animals and transferred into old mice, this did not occur. In other words, it was not the ovary or the eggs that had irreversibly aged, but the chemical (or hormonal) environment that had apparently changed. Although it is difficult to generalise these findings to other tissues of the body, it nevertheless indicates that at least some cells cease functioning with aging, not because their Hayflick limit runs out – but because the chemical milieu surrounding them changes.

Another type of tissue that has been used for transplantation is mammary tissue. In mice, the mammary glands begin to form shortly after birth with the development of small bud of cells (localised around the nipple) that form branching tubules which infiltrate into the surrounding fatty tissue. Within 6 to 8 weeks of birth the mammary glands become fully formed. Charles Daniel, working at the University of California, has shown that these small mammary buds can be surgically removed and placed into new hosts where, they will develop into fully functional tissue capable of producing milk if their hosts become pregnant. Importantly, the buds of cells that form the gland can again be removed from the host and placed into another mouse where they will form new tissue. Thus, this situation provides a good opportunity to see whether this tissue can repeat the cell proliferation cycle indefinitely, or whether a limit to replication occurs as predicted by Hayflick.

Daniel found that when mammary gland cells were serially transplanted at regular 3 month intervals, the amount of cellular proliferation as measured by the percentage of available fat filled by the transplant, declined in a regular fashion. After 6 transplantations, the amount of fat tissue innervated by new cells had declined by 80%, and by the 7th operation the procedure had effectively stopped working. In other words, it appeared that these cells had reached the end of their Hayflick limit. A second important issue concerned the viability of transplanting tissue from old donors. To examine this, Daniel tested the performance of cell grafts taken from 3 week and 26 month old mice, that were implanted into fat pads on opposite sides of the abdomen in new hosts. The results showed that there was no difference between the two groups in the number of serial transplants that could successfully be made. Therefore, these results suggest that mammary epithelial cells do not 'age' over a normal mouse life span, but rather their ability to divide becomes limited when serially transplanted. In one respect this appears to support Hayflick's theory that it is the number of cellular divisions, and not chronological time, that is the crucial factor in determining the life span of the cell. However, in another respect, it also shows that the divisional potential of the mammary cells do not correlate very well with the life span of the individual.

What would happen if cells were not restrained by their Hayflick limit? Would cells make copies of themselves *ad infinitum* and be immortal? The answer, as incredible as it may seem, is yes. In short, there are some cells that can be taken from the body and kept alive in tissue culture indefinitely. Unfortunately, these are not cells that we would want to be personally associated with because they are abnormal and the cause of one of our biggest killers – cancer. Many different types of cancerous cell have been cultured, but the oldest and most famous are HeLa cells that were taken from a women called Henrietta Lacks who died from cancer of the cervix in the early 1950s. Samples of her cancer cells were transferred to a culture medium by George Gey at John Hopkins University in 1952, and they have continued to thrive and replicate ever since. Unlike normal cells, HeLa cells have over-ridden their Hayflick limit, and do not respond to the normal signals that tell them to stop dividing. Consequently, when placed into a culture medium they vigorously divide and pile up on each other in a disorganised mess. Their immortality is confirmed by the fact that these cells have been distributed to laboratories all around the world where they still continue to replicate (if given nutrition) without showing the slightest indication of slowing down.

Cancer is probably the most feared disease of our time. At present it is second only to heart disease as the leading cause of death in the Western world, and within a few more years it is probable that cancer will overtake the latter to become the number one killer. It is essentially a disorder of uncontrolled cell growth of otherwise healthy cells. Thus, cancer cells continue to grow and divide without restraint, forming a tumour (or 'neoplasm') which may, if malignant, spread throughout the body, interfering with the function of normal tissue and organs and progressively leading to death. At first sight it would appear that the cancerous cell has lost the intrinsic (genetic) braking mechanism that comes into play when the Hayflick limit has been reached. In fact, things are a lot more complex than this, although as we shall see below, an understanding of cancer may have implications for understanding the underlying causes of the Hayflick limit.

To understand cancer more fully it is important to remember that the normal (or *controlled*) process of cell division is an essential part of body growth and tissue repair. For example, the division of cells have to take place in growing children to fulfil a predetermined pattern of development; and during adulthood our cells retain this capacity so that they can renew worn-out or damaged tissue. Indeed, the loss and renewal of cells is a constant process and operates on a huge scale. An average adult consists of about three trillion cells and it has been estimated that about 350 billion of these divide every day. For example, we shed millions of cells from our skin; the lining of the intestines; and from the insides of our airways. In addition, vast numbers of old red and white blood cells are continually being replaced by new cells every moment of the day. These processes also have to be finely tuned and synchronised so that cell renewal matches cell loss (this suggests that complex genetic control systems are involved). Unfortunately, it only takes one rogue cell to disrupt this delicate balance and for the whole system to come under threat, and this is effectively what the cancer cell manages to do.

It is a frightening fact, but cancer nearly always arises from one single abnormal cell. Since each cancerous cell will divide to form two new cancer cells, a quick calculation shows that the total number of new cells can increase dramatically in a very short time. As few as 20 cell divisions will result in the formation of about one million cancer cells, and yet another 20 cell divisions will increase the number of cells in a hypothetical tumour to about 1 trillion – which roughly corresponds to about a pound of tissue. In fact, if the tumour grows under optimal conditions then this amount of tissue could form in as little as 40 days. Although this type of situation is highly unusual, it nevertheless illustrates just how dangerous to the body one single cancerous cell can be.

One of the biggest breakthroughs in the last 20 years or so, has been the identification of certain genes whose damage is responsible for the development of cancer. There are, in fact, two groups of genes that are known to be involved in the formation of a cancerous cell: the proto-oncogenes ("onco" means cancer); and the tumour suppressor genes. The first of these (proto-oncogenes), are normal cell genes that are found in all vertebrates, and whose role is to regulate the many aspects of cellular growth, division and differentiation. Although these are normal genes, they can nevertheless suffer damage and be transformed into oncogenes which causes the cell to divide abnormally. In fact, around 60 proto-oncogenes have been identified and all of them can be converted into an oncogene that plays a dominant role in the formation of human cancer. This high number of proto-oncogenes indicates that the cell is surprisingly vulnerable to oncogene formation, and probably because of this, it has evolved a second set of genes whose function it is to produce proteins that suppress inappropriate growth. Some of these genes appear to become switched on when the cell is under high threat (e.g., high radiation exposure) whilst others seem to be continually at work holding back cellular growth. An example of a tumour suppression gene is the p53 gene that has already been mentioned in relation to apoptosis (chapter 4). Over half of all human malignancies, including those of the lung, colon and breast, have been shown to have damage to this gene, making it a huge risk factor for cancer.

Although cancer arises from a single cell, the process by which it is transformed into a cancerous cell is a slow and gradual process. Normal cells do not become converted into cancer cells overnight, or in a single mutational step. In fact, for a cancerous cell to develop, mutations must occur in at least two genes, and in many cases, in at least half a dozen. This can be shown by the fact that there is nearly always a long delay between the causal event and the onset of the disease. For example, the incidence of lung cancer does not increase sharply until some 10 or 20 years of heavy smoking; and the incidence of leukemias in Hiroshima and Nagasaki did not show a marked rise until about 5 years after the explosion of the atomic bomb. Indeed, it takes considerable time for genetic mutations to accumulate and, moreover, the fact that human cells have considerable DNA repair capacities, have the benefits of tumour suppressor genes, and even then has the back-up of apoptosis if anything goes wrong, all contrive to make the emergence of a cancerous cell an unlikely event.

As we have seen, a cancerous cell is essentially one that has broken free from the 'stop' signals telling it to cease replication. The Hayflick limit would appear to represent the other side of the coin to cancer since it is a cellular state where the cell has been put into a permanent state of non-division. This suggests that there may be an underlying common genetic link to both cancer and the Hayflick limit. And perhaps more importantly, if we understand the mechanisms of one, this may well have implications for understanding the other.

THEORIES OF THE HAYFLICK LIMIT

If we accept that the Hayflick limit is relevant to the aging process (although this is by no means certain – see pages 152–155) then clearly the fibroblast technique provides us with a very simple and useful model of aging 'in miniature' by which to examine its *causes*. It follows from this that if we can understand the reasons why cells stop proliferating, then we may well obtain a valuable, if not crucial, insight into the reasons why we age. It is not surprising therefore, that the quest to understand the underlying biological mechanisms that cause fibroblasts (and other cells) to stop dividing and die is one of the most actively researched (and controversial) areas in gerontology today. This research has led to the development of two different types of explanation that can be called stochastic and genetic. The first proposes that cells reach their Hayflick limit because they slowly accumulate damage resulting from wear and tear which eventually interferes with their ability to replicate. This damage may include somatic mutations, structural faults in proteins, or the accumulation of intracellular waste products. In contrast, the second explanation holds that the limited ability of cells to divide is programed into its DNA. This implies that our cells contain a counting device that is able to record how many times it has divided, and then once a set total has been reached, a second mechanism comes in to play that turns-off further division. As we have seen in previous chapters, these two theories lie at the heart of the debate on the causes of aging. Understanding the causes of the Hayflick limit would, therefore, appear to offer us a simple means of assessing the relative merits of these two explanations.

GENETIC THEORIES

We have encountered evidence supporting the genetic hypothesis with Hayflick's own work showing that it is the nucleus (rather than the cytoplasm) that is the main site which determines the number of times a cell can divide. Thus, it seems highly probable that the Hayflick limit must have a genetic basis. But where do we begin to explain what happens? In one sense the gist of the answer appears to be straightforward. Presumably, there are sets of genes that control the division of the cell (indeed, as we have seen, a large number of these genes are known from cancer research) and at some point, these genes are closed down and no longer able to be activated. This is an attractive idea because a similar process is known to underlie cellular differentiation (i.e., the process by which some cells become skin cells, and others become liver cells, despite having exactly the same genetic material), and moreover it is clear that any given cell must include sets of genes that are either turned on or off. The problem, therefore, is trying to explain how this situation comes about.

The answer to this problem was first supplied by the Nobel Prize winning French biologists Jaques Monod and Francois Jacob in 1961. As we saw in the previous chapter, genes are only expressed when they are transcribed into mRNA (so that the gene's message can be transported to a ribosome) by an enzyme called RNA polymerase. This enzyme sits in place near the start of the gene (in an area known as the promotor region), primed for action and ready to move down the

DNA. However, in genes that are switched-off, the RNA polymerase is effectively put out of action by a part of the DNA known as an operator, which acts as a barrier to the enzyme's progress. In other words, the RNA polymerase is literally blocked by the operator from moving down and transcribing the gene. The operator therefore acts as a switch, but to enable it to serve this role it also requires the help of a special repressor protein (which is made by another 'regulatory' gene). In short, when the operator and regulatory protein are bound together, they act to inhibit RNA polymerase activity, and it is only when the operator loses the help of its repressor protein, that the RNA polymerase is free to begin its journey down the DNA (see figure 6.3).

Could it be that the Hayflick limit is due to certain genes normally involved in cellular division being turned-off by the formation of special repressor proteins that attach themselves to their operons? Evidence for this idea has come from cell fusion experiments in which cells are fused together to form a cell with combined cytoplasm and two separate nuclei (this is slightly different to Hayflick's work described on pages 139–140). Early work using this procedure found that if senescent cells (that had reached their Hayflick limit) were fused with young cells, then

(A) With repressor protein in place, RNA polymerase cannot move down the gene, in effect turning the gene off.

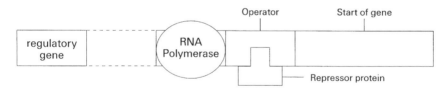

(B) With the repressor protein removed, RNA polymerase can now transcribe the gene, in effect turning it on.

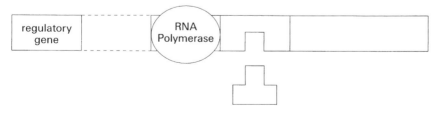

Figure 6.3 Diagram showing simplified activity of an operon and its control of gene transcription.

the 'young' nucleus quickly became inhibited and unable to synthesise DNA. Similarly, when cells of different ages (young and old) were fused together, the replicative potential of the fused cell more closely resembles that of the older cell than the younger one. The simplest interpretation of these results is that the old cells produce a substance that blocks the DNA synthesis in the young nucleus. Although this substance has yet to be formally identified (it is presumably a protein), this type of research suggests that the genes do not simply fail, but rather they are actively shut down by a substance that is made by the cell itself. Indeed, one can imagine this substance building up with each cell division, eventually reaching a level where it is able to shut down the genes controlling cellular division.

An interesting variation of this type of experiment has examined the effects of fusing normal healthy cells with cells that have been made immortal by viral infection. For example, when senescent fibroblast cells are fused with 'immortal' HeLa cells, it is found that DNA synthesis is reinstated in the 'old' senescent nucleus. Thus, it would appear that HeLa cells contain a substance that overrides the inhibitory signal that is telling the old nucleus to stop dividing. However, when these rejuvenated old cells begin to divide and form new hybrid cells, they show a limited ability to divide and typically go through fewer than 8 replications. The rejuvenation is therefore short-lived – suggesting that the genes responsible for shutting-down cell division are more dominant in the long run than the newly introduced genes producing unlimited replication.

Some of the most compelling evidence for a genetic involvement in limiting cellular division has come from work showing that the introduction of certain individual chromosomes into immortal tumour cells can actually cause them to stop dividing and become senescent. So far this has been achieved with the incorporation of human chromosomes 1 and 4 into human tumour cells, which indicates that these chromosomes may have a particularly important role in directing cellular senescence. Other lines of research have examined the expression of certain genes in senescent fibroblast cells. For example, the husband and wife team James Smith and Olivia Periera-Smith have found that certain genes are expressed (i.e., become more active) when the cells approach their last division; while others (including the p53 gene) become less active. It remains to be seen whether these genes are directly involved in controlling the Hayflick limit, but the participation of at least some of them appears likely.

THE TELOMERE THEORY

There is, however, an alternative genetic theory that does not depend upon inhibitory proteins and the expression of genes to explain the Hayflick limit, and this is known as the telomere loss theory. Telomeres are special sequences of DNA that are found at the ends of the chromosomes, and which have the function of stopping chromosomes fusing together. Although telomeres are made of DNA they do not contain any genes; rather they are made up of a large number of repeating subunits of bases (in humans the sequence is TTAGGG). Research by Calvin Harley at McMaster University in Ontario, Canada, has shown that with every cell division that takes place in normal cultured human fibroblast cells, the number of repeated sequences become less. In fact, this process continues until the telomere loss starts to cut into the bases of the genes needed by the cell, at which point the function of the chromosome becomes severely compromised. Not surprisingly, it has been suggested that telomere loss may function as a biological clock that counts the number of divisions the cell has made. The idea that progressive loss of telomere DNA may be the cause of aging was first suggested by A.M. Olovnikov in 1973, and the idea has continued to gain support. For example, it has been shown that the length of telomere DNA taken from human fibroblast cells correlates very well with the ages of the donors; and people with progeria have significantly shorter telomere lengths than normal people of the same age. In short, the length of telomeres is an excellent predictor of the replicative capacity of human fibroblast cells.

It is also believed that telomeres serve another important function by stabilising the chromosomes of which they are part. Without this stabilising influence, chromosomes might fuse together or break into segments that rejoin in abnormal combinations. Indeed, it is known that chromosomal damage increases with age, and it is possible that telomeres have a role to play in this effect. Although irrefutable evidence is hard to obtain, it would perhaps be surprising if these changes didn't contribute further to the decline of the cell and its senescence.

Interestingly, it has been shown that cancerous cells have found a way of stopping telomere loss by producing an enzyme called telomerase that rebuilds telomeres at each round of cell division. Most of our body cells do not contain this enzyme, but the germ cells provide an important exception (of course, this makes sense because our eggs and sperm need protection from the loss of telomeres to ensure that they are immortal in order to serve as the units of inheritance). Despite this, it appears that all our body (or soma) cells actually contains the gene that produces telomerase except that it is not switched on; but, for some unknown reason, this telomerase gene becomes activated during the development of cancer. Indeed, Calvin Harley has shown that if HeLa cells are deprived of telomerase they quickly die. This raises the exciting possibility that if drugs can be developed that inhibit telomerase activity, then a cure for cancer may perhaps be found.

Despite this, there are other findings that raise strong doubts about the telomere theory of aging. The most problematical is the fact that rodents have long lengths of telomeres that do not appear to shorten with aging. Furthermore, not only do they express telomerase in their germ cells, but they do so in their body cells as well. This finding indicates that the loss of telomeres is not a universal feature of animal life and, therefore, presumably is not a universal feature of aging. Clearly, if telomere shortening is an important cause of cellular decline and human aging, this raises the uncomfortable possibility that the underlying mechanisms of aging might be very different between species. Either that, or telomeres are not important after all.

STOCHASTIC THEORIES

It is fair to say that stochastic theories have been overshadowed in recent years by attempts to find a genetic explanation for the Hayflick limit. Nevertheless, it remains that fibroblasts are subject to wear and tear, and are susceptible to a wide range of potential damage (including that arising from somatic mutations, error catastrophe, and free radical reactions). Although the cell has defences to protect itself, and mechanisms to correct any damage that it incurs, no form of cellular protection is perfect, and it is possible that small amounts of genetic damage accumulate in the cell every time it replicates itself. Although this damage may be relatively minor, it might nevertheless gradually build up over numerous generations so that more and more defects will occur. If this is the case, then eventually the cell is likely to reach a situation where further division is no longer possible, or senescent changes begin to occur.

The evidence for this type of stochastic progression doesn't appear to be particularly strong although it cannot be ruled out. One way of testing the theory is to expose fibroblast cells to ultraviolet radiation or certain chemicals that are known to produce random genetic damage (somatic mutations). In theory, this should result in more genetic damage being passed on during the process of cellular replication, which in turn, might be expected to accumulate with each division. However, research has shown that increasing the amount of somatic mutations has relatively little effect on inhibiting fibroblast cells from dividing, or reaching their Hayflick limit. In fact, it has been shown that mutation rates in fibroblast cells have to be some 100-fold higher than normal before they have any real effect on cellular senescence. One reason for this somewhat surprising finding is that fibroblast cells have very efficient levels of DNA repair and, it appears that this capacity does not decline until the cell begins to reach the end of its replicative life. Thus, the fibroblast cell seems to be reasonably well protected against the damaging effects of somatic mutations.

An alternative theory of why fibroblast cells age is that they accumulate a number of protein errors; the most important of which are the enzymes that are involved in the transcription and translation of DNA that, could in turn, lead to error catastrophe (see previous chapter). Indeed, evidence supporting this theory has come from research showing that the fidelity of DNA polymerase (which is the enzyme responsible for DNA synthesis) decreases with the number of cellular divisions. If this is the case then one would predict a build-up of faulty proteins in late-dividing fibroblast cells. However, although biochemical research has found a number of enzymes that appear to be changed in aged fibroblast cells, there is little evidence that these cells go through an error catastrophe as might be expected. In fact, fibroblast cells can be induced to make a large number of faulty proteins by the introduction of amino acid analogues into their culture (see page 112) and these have been shown to have no effect on their ability to reach their Hayflick limit. Thus the evidence seems to indicate that whilst some proteins may be changed in aged fibroblast cells, they have relatively little effect on the fibroblast's ability to divide.

Surprisingly, there has been relatively little research examining the effects of free radical damage on the aging of fibroblast cells. An early study published in 1974 by Lester Packer and James Smith caused great excitement when they reported that the addition of the antioxidant vitamin E to fibroblast cultures could double their replicative potential to such an extent that they could go through 100 or more doublings without showing any signs of abnormality. However, this finding was not been repeated by other investigators, and has since been retracted by Packer and Smith. Another way of testing the effects of free radicals on the proliferative decline of fibroblasts is to grow them in reduced concentrations of oxygen, which might be expected to cut down the number of free radical reactions taking place in the cell and thus prolong their ability to divide. In the main this prediction has not been confirmed and, in short, the role of free radicals in the aging of fibroblast cells remains uncertain.

IS THE HAYFLICK LIMIT RELEVANT TO AGING?

This might seem an odd question. In spite of all the evidence presented so far, it might come as a surprise to find out that the relevance of the Hayflick limit to the understanding of aging remains highly controversial. One area of doubt centres on the issue of whether fibroblast cells that have reached their Hayflick limit are actually old and senescent (as Hayflick maintains), or whether they have simply changed their function and become a new type of cell. This latter idea is not as extreme as it first seems because there are many instances where cells change from unspecialised cells, into specialised ones, in a process known as differentiation. The most obvious example of this occurs just after conception, when the fertilised egg starts to differentiate into a wide variety of cells to form the developing embryo. But differentiation also occurs on a large scale in the adult. For example, all of the cells making up our blood and immune system are derived from the same basic cells located in the bone marrow. These are basically primitive cells (also known as stem cells) that do not look like any other type of cell, and have no real function except to give birth to other cells. And the way in which they do this is quite unique. Although most cells in the body give rise to two identical daughter cells when they divide, bone marrow stem cells undergo asymmetric division. That is, while one daughter is an exact replica of the mother cell (thus the cell is self-renewing), the other daughter becomes slightly more differentiated and starts on the pathway to becoming a much more specialised type of cell. In some instances this cell may differentiate further in the bone marrow (as do the red blood cells) or they may become more specialised after they have left the bone marrow (as do the white blood cells).

There are two important points to bear in mind regarding the division of stem cells in the bone marrow. The first is that the stem cells are effectively immortal since every time they divide they make a brand new copy of themselves. In other words, stem cells divide without limit, or at least they do so for the entire lifetime of the animal. However, once a daughter cell embarks on the alternative course then the spell becomes broken. These cells begin their journey to becoming more specialised, and in doing so they eventually reach their final mature stage where they become terminally differentiated. In contrast to the stem cells, these differentiated cells are mortal, they cannot reproduce themselves and when they die they are quickly replaced by new cells.

Returning to the arguments surrounding the Hayflick limit, it has been argued by some researchers that when cultured fibroblasts reach the end of their ability to divide, they are not senescent but have simply differentiated into a different type of cell. There is some justification for this viewpoint because in some respects, fibroblasts resemble undifferentiated cells. For example, in the body, fibroblast cells are essentially secretory cells, that in response to situations such as development or tissue damage, secrete a number of proteins (including collagen, reticular fibres and elastic fibres), as well as extracellular gel-like substances that help bond these fibres together. It can be argued that once fibroblast cells have served this secretory function they go on to become a new type of cell with a different function. Indeed, evidence for this position has come from studies that show that non-dividing fibroblast cells are not necessarily in terminal decline as Hayflick first believed. For example, by using appropriate culture techniques it has been possible to keep 'old' non-dividing fibroblast cells alive for over a year (i.e., in effect doubling their life span) during which time they appear to be perfectly normal. In other words, given the right conditions, fibroblasts do not simply stop dividing and die in the manner first suggested by Hayflick. The doubling of fibroblast life span in this way does not make a great deal of sense if these cells are seen as senescent, but it does make sense if one views these cells as fulfilling a new function.

Further evidence supporting this view has come from the work of Klaus Bayreuther and his colleagues at the University of Stuttgart, Germany, who have shown that during the fibroblast's life, the cells pass through seven different stages. Three of these stages are similar to the ones that Hayflick described, but the remaining four stages occur once the cell has actually stopped dividing. Not only is the morphology of the cell changed at each stage, but perhaps more importantly, different proteins are also made at each stage implying that each stage is under genetic control. Again, these findings do not make a great deal of sense if one assumes that old fibroblast cells are past their sell-by date, and serving no useful function.

A further variation of the differentiation theory was developed by Thomas Kirkwood and Robin Holliday in 1975. These researchers suggested that when embryonic fibroblasts are initially placed in culture they are potentially immortal (or in their terminology "uncommitted"). But as they divide, a small percentage of cells (which they estimate at about 0.275% per generation) differentiate and become 'committed' to some particular function. In doing this they change from immortal to mortal cells. Consequently, over time there will be a gradual accumulation of mortal cells in the culture medium. But, Kirkwood and Holiday also point out that a small percentage of immortal cells should always exist in the culture even after many divisions. The reason that this does not "appear" to happen, they argue, is that it is necessary when adopting Hayflick's culture technique, to discard most of the cells since to keep all the doublings would require literally millions of flasks. The consequence, according to Kirkwood and Holliday, is that most of the immortal cells become discarded, leaving behind the only type of cells that can be detected – namely the mortal cells. Viewed in this light, the Hayflick limit is seen as an artefact of the experimental technique that accidentally phases out immortal cells.

The differentiation theory in all its various forms has attracted a considerable amount of support. Unfortunately, its major stumbling block is that nobody has ever shown that a fibroblast cell is able to make a differentiated cell with a clearly defined specialised function. Or, more specifically, nobody has ever shown (or recognised) a late passage or 'aged' fibroblast cell to have any use or real function in the skin (or any other tissue). Consequently, until somebody actually discovers a differentiated cell with a specialised role that is derived from, and looks like, an 'aged' fibroblast cell, then there will always be some doubt concerning the theory.

Even if we dismiss the differentiation theory and accept that Hayflick is correct in his interpretation of what is happening to his fibroblasts, it is still not clear whether it tells us anything about aging. As we have already seen, not all cells in the body are capable of replication like fibroblasts. For example, a number of important cells including nerve cells that make up the brain and spinal cord, and heart muscle cells, do not divide once they are fully formed. Yet, these cells clearly age. Therefore, the Hayflick limit cannot explain how every cell in the body ages. Secondly, it is clear that living creatures do not die because they exceed their Hayflick limit, and even the oldest individuals appear to have reserves of fibroblasts to keep them going. To explain this apparent discrepancy, Hayflick argues that a declining loss of replication will result in cell loss that, in turn, will produce a decrement in the structure and function of various organs of the body. But, again, there is uncertainty over whether this really occurs. For one thing, the reserve of cells in various organs is enormous and secondly, there is little strong evidence showing that organs of the body decline because their cells are losing their ability to proliferate. Thus, despite all its attractive features, and its ability to excite the imagination, establishing the true relevance of the Hayflick limit to understanding aging remains highly elusive.

SUMMARY

- Many cells in the body are *mitotic* meaning that they are capable of division (or replication) although others such as brain and muscle cells are *post-mitotic* meaning that they are incapable of replacing themselves.
- Alexis Carrel managed to keep cultured cells alive for over 34 years and claimed that cells were essentially immortal (although he was mistaken). This view suggested that the cause of aging was *extrinsic* to the cell.
- In 1961, Hayflick and Moorhead showed that human fibroblast cells have a limited (50 ± 10) capacity to divide (now known as the 'Hayflick limit') which indicated that the cause of aging was *intrinsic* to the cell.
- Cells taken from older individuals tend to produce fewer replications than cells taken from younger individuals (e.g., George Martin has shown that there is a decrease of 0.2 divisions for each age of the donors age between the ages of 10 and 90). Moreover, longer-lived species tend to have a higher Hayflick limit than shorter-lived species.
- Hayflick showed that the mechanism that counted the number of cellular divisions appeared to lie in the nucleus (and presumably therefore in the genetic material) and not in the cytoplasm.
- Research examining tissue transplantation has provided some limited support for the idea that the Hayflick limit is an important component of aging. However, such work has also indicated that the external (neural and chemical) environment of the body is also an important factor in the maintenance of cells.
- Cancer cells are those that do not respond to signals telling them to stop dividing. Two classes of gene are known to be involved in the development of a cancerous cell (proto-oncogenes and tumour suppressor genes) and normally a combination of these genes have to be damaged for a cancer cell to arise.
- One possible explanation for why the Hayflick limit occurs is that cells produce a protein that builds up with each division and which eventually switches off the genes involved in replication.
- An alternative genetic theory holds that the Hayflick limit is due to the progressive loss of telomeres (located at the ends of chromosomes) which takes place every time the cell divides. In support of this idea

it has been shown that the length of telomere DNA taken from human cells correlates very well with the age of the donors.

- The relevance of the Hayflick limit to understanding aging remains controversial. For example, some researchers believe that once a cell has reached its Hayflick limit it has simply changed into a new type of cell. In addition, it seems unlikely that people age and die because their cells reach the end of their capacity to divide. Finally, the Hayflick limit cannot explain why cells that are unable to divide (i.e., post-mitotic cells) still continue to age.

METABOLISM AND AGING

In short, the length of life depends inversely on the rate of living

–Raymond Pearl

A man with a watch knows what time it is;
a man with two watches isn't so sure

–Anonymous

Energy is at the root of all life and the driving force underlying all the chemical reactions of the body. Each and every cell in the body needs energy to maintain itself, and the more work a cell does the more energy it will require. Most of this energy is used in the process of synthesising new biological molecules such as proteins and fats (anabolism), and in the breaking down of these large molecules into smaller ones (catabolism). The process of building up and breaking down materials is called metabolism, and the body's overall total energy expenditure per given time unit, is called the metabolic rate. The concept of metabolism has always been particularly relevant to the issue of aging. On an individual level, changes in metabolism (or rather changes in the efficiency of how energy is used across the life span) have important implications for understanding the decline of physiological function that occurs with aging. But the concept of metabolism also provides a much broader perspective by which to understand the aging process. In 1918, the German biologist Max Rubner made the remarkable observation that all animals appear to use up the same amount of energy during their life. For example, a mouse and an elephant, despite their greatly different life spans, will actually consume a similar amount of energy per gram (or kilogram) of their body mass during their life. Put another way, if one could somehow take a equal amount of mouse and elephant tissue, one would find that despite the fact that the mouse expends its energy much more quickly than the elephant (and presumably because of this lives for a shorter period of time), the amount of energy used-up by both units of tissue over their life would be very similar. This is much the same as saying that animals that burn-up energy faster die quicker. This relationship has continued to fascinate gerontologists for decades and has provided important insights into the biological basis of aging.

ENERGY AND METABOLISM

Energy is not an easy concept to understand. It is not a substance that has a size, shape or mass, but rather it is something that can only be inferred and measured when some type of change is occurring. Put another way, energy can be defined as the capacity to do work, or to put matter into motion. In living organisms, energy is stored in chemical bonds that hold atoms and molecules together, and is only released during chemical reactions when energy is transferred from molecule to molecule. Thus, energy occurs when molecules are moved about by the realignment of their atomic particles, and one consequence of this movement is the production of heat. One way of understanding energy, therefore, is to see it as the potential to make molecular (or cellular) movement and heat.

Another way of understanding energy in living organisms is in terms of oxygen consumption. The main source of energy for cells is glucose, but the cell is unable to use this substance directly for its energy needs and, consequently, has to transform it through a series of complex chemical steps involving oxygen into a molecule known as adenosine triphosphate (ATP). This is the all-important substance that stores and releases chemical energy for cellular use in the body – and without it life would quickly cease. In fact, ATP can be likened to a coiled spring which is able to uncoil with tremendous energy when its catch is released. In this way it is able to transfer its energy to other molecules to drive biological processes. To create ATP, however, the cell needs oxygen. Thus, if one wants to measure the rate of energy expenditure (that is the formation and use of ATP) one can also go about it by measuring oxygen consumption.

Thus, the total energy expenditure (or metabolic rate) of any animal can be worked out by measuring the amount of energy that is lost by the body as heat, or by measuring the body's oxygen consumption. Indeed, biologists have long known that oxygen consumption and heat production are directly proportional. For example, for each litre of oxygen used by the human body, a total of 4.8 kcalories of heat is generated (a calorie is the quantity of heat that is required to raise the temperature of 1 gram of water by 1°C).

So, if we want to measure the metabolic rate which, in effect, is the total sum of all chemical reactions taking place in the body, all we need to do is to accurately measure its heat loss or oxygen consumption.

THE BASAL METABOLIC RATE

There are several techniques that we can use to measure the metabolic rate: perhaps the simplest is to use a respirometer (which measures the inhalation of oxygen), and then to multiply this amount by 4.8 (the amount of energy in kcals liberated by 1 litre of oxygen). This value provides an estimate of the rate at which fuel is being utilised by the body to provide energy for all of its activities. However, it is apparent that the rate at which metabolism occurs will depend upon how quickly a person is using up their energy (i.e., a resting person will have a much lower metabolic rate than an active one.) To allow meaningful comparisons of the metabolic rate to be made between individuals therefore, it is necessary to test subjects under standardised conditions. This is normally undertaken by getting subjects to fast for at least 12 hours; to be mentally and physically relaxed; and to be adapted to a room temperature of 20–25°C. The amount of oxygen used under these conditions provides what is known as the basal metabolic rate (BMR), or what is sometimes referred to as the 'energy cost of living'. In this resting state it has been shown that humans need to consume approximately 200–250 mls of oxygen per minute, which is sufficient to allow an average 70 kg (154 pound) adult to use up approximately 60–72 calories of energy per hour. In effect, this is essentially the minimum amount of energy needed to maintain life.

The basal metabolic rate provides energy for many different functions. The largest structural component of the human body is the muscle tissue which makes-up nearly 50% of the body's mass. During the body's resting state, muscular activity is needed to maintain circulation and respiration, as well as providing body tone: and these functions take-up some 20% of the body's energy expenditure. The liver also requires a high basal expenditure of energy (accounting for about 20% of the BMR), as does the brain which although only accounting for about 2% of the body weight, nevertheless consumes a remarkable 20% of the body's energy supply (the brain's high energy demands can be shown by the fact that if it is starved of oxygen for only 10–15 seconds it will enter a state of unconsciousness). Much of the remainder of the basal metabolism is used to maintain and replace cells throughout the body, including those in the skin, immune system, blood and so on.

A number of factors influence the rate of basal metabolism in humans. One factor already mentioned is muscle mass. Because muscle even in its resting state uses up considerable energy, individuals with more muscle mass will tend to have a higher basal metabolism. However, an even more important determinant of the basal metabolic rate is the surface area of the body. This is because as the ratio of body surface to body volume increases, heat loss to the environment also increases, and therefore the metabolic rate has to be higher to replace the lost heat. Consequently, if two people have the same weight, then the taller or thinner person on average should have the higher basal metabolic rate compared to the shorter or fatter person, because they are less able to conserve heat. There are also many other factors that can influence the BMR including gender, external temperature, stress, hormone levels (particularly those of the thyroid gland), and food consumption (e.g., the BMR can increase by as much as 20% after a meal).

THE EFFECTS OF AGE ON METABOLISM

Most studies have shown that the amount of oxygen, or energy, necessary to maintain the resting state of the body declines with age. On average, the human body reduces its energy expenditure by about 12 calories per year after the age of 30, which represents a decline of 120 calories every decade. Thus, on first sight, it appears that basal metabolism declines significantly with advancing age. However, this decline in the body's energy needs is not as simple as it first appears because changes also take place in the composition of the body with aging. More specifically, aging is usually accompanied by a decrease in muscle mass along with an increase in fat deposits, and as we have seen, muscle has much higher metabolic requirements than fat. For this reason alone we might expect a decrease in the body's energy requirements to take place with aging. Indeed, if a correction is made for the decline in muscle loss, then it is found that the basal metabolic rate does not change significantly with aging. In other words, the body's tissues require the same amount of energy to maintain their resting function in young and old alike. This is further supported by the finding that there is no age difference for mouth or axillary temperatures taken from individuals ranging in age from 20 to 100 years. In short, the basal production of body temperature does not appear to change with age.

On average, protein (or lean body mass) declines at a rate of about 6 pounds per decade after early adulthood, with this figure tending to decrease further after the age of 50 years. Consequently, by the time men reach 70 years they have typically lost about 20 pounds of muscle mass; and in compensation gained some 7.5 pounds of fat and connective tissue, compared to when they were 40. In some cases muscle loss may be far greater, and it is not unusual for octogenarians to have lost a third of their muscular mass from early adulthood. Many researchers believe, however, that this age-related loss is not inevitable, and is largely due to lack of exercise since adults who remain physically active as they age retain much more of their muscle mass and consequently show a much smaller decline in basal metabolism. It might be more accurate to say, therefore, that the decline of energy expenditure with aging is more to do with the decline in physical activity than any underlying change in metabolism.

Although there does not appear to be any age-related decline in the body's energy requirements, there is a decline in the body's efficiency to use energy over and above the resting levels. In other words, it usually takes more energy for an older person to do a given piece of work than a younger person, especially if a task is strenuous. This can be shown in a number of ways. For example, moderate to heavy exercise normally requires much higher oxygen consumption along with greater increases in heart rate in elderly subjects. Alternatively, a study undertaken in 1950 found that when young and old men were exposed naked to prolonged cold temperatures, the aged subjects were less able to maintain their normal body temperature and used up more oxygen in the process. Indeed, if one measures the rate of oxygen and carbon dioxide that is exhaled in the breath of elderly subjects, one finds that the proportion of oxygen increases whilst the level of carbon dioxide decreases. In other words, it appears that the aged tissues of the body cannot utilise oxygen at the same rate as younger tissue.

THE ROLE OF MITOCHONDRIA

If a decline in the efficiency of metabolism takes place with aging, why then does this change comes about? One explanation lies with small rod-like structures that are found in all energy creating cells of the body called mitochondria. These tiny structures (which can only be seen with the aid of an electron microscope), are the cell's power stations which provide the energy to drive all the chemical reactions necessary for life. Most of our cells contain hundreds or even thousands of mitochondria, and they are often likened to small batteries without which the cell would have no supply of energy and quickly die. It is believed that mitochondria were once free-living bacteria, but were 'captured' millions of years ago by more advanced eukaryotic cells (cells which contain a nucleus) which then began to use them for their own energy needs. This symbiotic relationship evolved, however, so that now the cell and mitochondria cannot exist without each other.

Indeed, mitochondria still have much in common with their ancestral relatives, including owning their own genetic material which is arranged in a ring (and not in a chromosome like that found in other bacteria). Moreover, because of their genetic inheritance, mitochondria have the ability to make their own proteins, and they are also able to independently replicate themselves *within* the cell. It is also interesting to note that we inherit all our mitochondrial genes from our mother, unlike our chromosomal DNA which is derived from both parents.

The process by which mitochondria are able to produce energy for the cell is extremely complex and includes a long sequence of chemical reactions involving many different types of enzymes. But, put simply, our mitochondria take chemical energy ultimately derived from food molecules and, with the aid of oxygen (which tends to free carbon atoms in the process thus making carbon dioxide) transforms these molecules into ATP (see above), which then diffuses throughout the cell where it is used to drive a wide variety of chemical reactions. Mitochondria therefore serve a vital function: without these structures the cells would have insufficient energy to maintain their existence. Indeed, this is basically what happens in cyanide poisoning which blocks the action of an enzyme (called cytochrome oxidase) in the mitochondria, which stops respiration and quickly depletes the cell of ATP.

MITOCHONDRIA AND AGING

Mitochondria are the powerhouses that provide the cell with its energy, and it follows that any decline in the efficiency of mitochondria is likely to have potentially serious implications for the cell's capacity to carry out its work. But why should a decline in mitochondrial function come about? There is one very good reason and that is: they are the main site in the cell where the formation of free radicals take place. These oxygen-derived chemicals, as we have seen in previous chapters, cause considerable damage to biological tissues and it has been estimated that around 90% of all free radicals produced by the cell are formed in the mitochondria. Thus, the mitochondria are continually producing highly reactive agents with the potential to cause themselves injury and deterioration. For this reason we might expect significant mitochondrial damage and a decline in their number with aging.

It might come as a surprise, but much of the work examining the role of mitochondria in aging has come from research involving insects. This is because their flight muscles require extremely high amounts of energy to perform their work, and consequently these muscles contain high numbers of mitochondria. For example, the wings of a fruit fly can beat at a frequency of some 300 cycles per second, and the wings of the common blow fly beat at around 180 cycles per second (both of which are far greater than say a hummingbird which only has a wing beat of about 70 beats per second). It is hardly surprising, therefore, that there can be a 100 fold difference in the rate of oxygen consumption between the resting and flying states of insect wing muscle; and to cope with these burst of activity the muscle requires plenty of mitochondria. Indeed, not only are these mitochondria relatively large but they comprise some 40% of the total muscle mass. Thus, the effects of exertion and aging on mitochondria located in insect flight muscle can be easily observed, at least in comparison to other species that have smaller and fewer numbers of mitochondria.

The strategy often adopted by investigators to assess the performance of insect flight muscle is to make insects fly until they drop exhausted. Using the fruit fly (*drosophilia funebris*) it has been found that flight performance reaches a peak around 7 days of age when the average length of flight is 110 minutes, and the total number of wing beats during this period exceeds 1 million. In flies that are 33 days old and that have nearly reached the end of their life span, however, the average flight time falls to 19 minutes and the number of wing beats declines to 170,000. Similarly, in the male housefly the duration of sustained flight has been shown to decline from 420 minutes at 1 day of age, to 63 minutes at 9 days, and a continual decline thereafter so that old houseflies eventually become unable to fly. In other words, this is a very marked decline in flight performance in a very short period of time.

Accompanying the decline in performance are changes in the size, number and shape of mitochondria taken from flight muscle. The first change that appears to take place in the housefly is the fusing together of mitochondria that starts at around 2 days of age. Over the next 7 days of the fly's life, the number of mitochondria decreases by 44%, although they increase in size by 143%. After this age it appears that the mitochondria start to become swollen and show signs of degeneration, and this is reflected in their performance on several measures. For example, oxygen consumption in houseflies has been shown to decline in a manner that closely corresponds to their overall activity; and furthermore there is a decline in mitochondrial efficiency as measured by their ability to produce ATP or to produce new protein. In short, the aging of insect flight muscle appears to be strongly linked to the degenerative decline of mitochondria.

What about the role of mitochondria in mammalian aging? Does an analogous decline of mitochondrial function occur in our tissues as in that of insects? Fortunately, the answer seems to be no, or at least the decline of mitochondria with aging has not been shown to be so dramatic and clear-cut. The big difference between insect flight muscle cells, and comparable non-dividing (post-mitotic) cells in mammals (such as muscle cells and neurons), is that the latter contain mitochondria that are able to renew themselves by self-replicating. So, whilst the numbers of mitochondria in insect flight muscle are fixed, the mitochondria in mammalian cells are often replaced before they are worn out. It has therefore been difficult to prove convincingly that mitochondria taken from young and old cells show any structural differences in mammalian tissue.

Mitochondria can replace themselves in this way because they contain their own DNA with which they are able to replicate themselves. However, unlike the DNA found in the cell's nucleus, the mitochondrial genome has no capability for DNA repair and, consequently, if any of its genes suffer a mutation, or any other kind of genetic damage, they remain unrepaired. The lack of any DNA repair system makes the mitochondria especially vulnerable since they are continually exposed to electron leakage and free radical damage. Consequently, it has been shown that the mitochondrial DNA accumulate mutations some 16 times faster than DNA found in the cell's nucleus. But, in one sense, this does not appear to be a big problem for many types of cell since they simply replace the damaged mitochondria with a totally new one. This is presumably one of the reasons why our muscle cells can last much longer than those in insects.

This is not to say, however, that our mitochondria have no role to play in aging. Indeed, as we have seen above, our aging cells do seem to be less efficient at using energy over and above basal resting levels. One likely explanation would seem to be a decline in mitochondrial function. Research by Jamie Miquel, working at the University of Alicante in Spain, has provided support for this idea by showing that some cells of the body contain fewer mitochondria as they age along with a resultant decline in ATP production. But not all cells show this change. As we saw in the previous chapter, some cells in the body (such as skin cells) are constantly replicating, whereas other 'post-mitotic' cells (such as brain cells) are unable to make new copies of themselves. Miquel has shown that the renewal of mitochondria in post-mitotic cells (and in cells with low replication rates) is much slower than that which takes place in rapidly dividing cells; consequently, non-dividing cells will tend to accumulate more damage thus impairing their function. This type of aging decline is therefore most likely affect the brain and nervous systems, as well as our skeletal muscle and cardiovascular systems. In contrast, the dividing cells of the body do not appear to undergo any significant degree of mitochondrial senescence.

NUTRITIONAL HOMEOSTASIS AND AGING

Although the basal metabolic rate, when adjusted to take into account changes in body composition, does not decline with aging, there are nevertheless many changes in the way the body controls and directs cellular metabolism. Among the many things that influence metabolic rate are levels of certain hormones (see next chapter) and activity in the autonomic nervous system (which regulates body functions not under conscious control such as heart rate, sweating, etc.). Both these systems change with aging. For example, circulating hormone levels tend to decline thereby reducing their influence on metabolism; and our cells show a blunted metabolic effect to norepinphrine despite it increasing significantly in the blood with aging. Similarly, the increase in metabolism that normally takes place following a meal is significantly reduced in the elderly. In short, whilst the basal metabolic rate does not change with aging, there are changes in the way it is controlled by the neural and hormonal systems of the body.

One of the most consistent features of aging, however, is a decline in the ability of the blood to restore normal levels of glucose following a meal. In young people, blood glucose levels return to normal after about two hours, but in older people this process takes much longer. In fact, by the age of 60, around 40% of people are so impaired in their ability to metabolise glucose that they could legitimately be diagnosed as diabetic if age-adjusted criteria were not employed. Indeed, the incidence of diabetes nearly doubles between the ages of 45 and 65 years, and it may occur in as many as 20% of people aged between 65 and 74. But even healthy aged people show a decline in their ability to handle glucose. For example, after the age of 30, peak glucose levels after a meal rise on average by 4 mg per ml every decade (probably due to a decline in the sensitivity of cells to use insulin), and there is also a small increase of around 1 mg per ml in fasting plasma levels with each decade of life. Thus, impaired glucose utilisation appears to be one of the many inevitable features of aging.

It is tempting to think that a reduction in glucose tolerance is a relatively unimportant contributor to the aging process, but there is increasing evidence to suggest otherwise. People with diabetes have a high likelihood of developing severe coronary heart disease, degeneration of the arteries and microcirculatory damage. In addition there is a risk of eye disease, neuropathy of the nerve systems (which can result in impaired movement co-ordination, incontinence, and decreased sweating), as well as joint stiffness and wasting. In fact, the effect of diabetes on many organs and tissues has often been described as accelerated aging, and this has led the biochemist Anthony Cerami, to propose that the age-related decline in glucose tolerance might also play a lesser but similar role in 'normal' aging as well. In support of this idea is evidence showing that glucose is able to react with proteins to form advanced glycation end products that are a major cause of cross-links in the body (see chapter 5). Glycosylation of proteins within the circulatory system, in particular, has been implicated in strokes and heart disease; and is an obvious risk associated with the increased blood glucose levels that, as we have seen, tend to accompany aging.

EVOLUTIONARY ASPECTS OF METABOLISM

There is another important way of looking at the subject of metabolism. Instead of seeing it as a cause of aging for the *individual*, we can also view it from an evolutionary viewpoint and regard it as an important determinant of *species* life span. For example, why does an elephant have a greater life span than a mouse? The most important reason is probably to do with their relative rates of metabolism. It goes without saying that a large animal needs more energy to keep its greater bulk alive than a small one and, of course, this is why an elephant consumes a much greater amount of food than a mouse. But if one compares the energy released per unit of body weight for the elephant and the mouse, one finds that the mouse spends its energy much more quickly. In fact, for each gram of body weight, the mouse consumes energy 20 times faster than the elephant – that is, if the elephant consumed energy at the same rate as the mouse it would require 20 times more energy just to keep alive. Viewed in this light it is hardly surprising that the mouse and elephant have such different life spans. But why do these two animals have different rates of metabolism? The most important reason (in this example) is their size.

Biologists have long known that as body size increases, the rate of metabolism decreases. To understand why, imagine we are able to double the height of a mouse and retain all its body proportions. We would find that this would increases the surface area of the mouse by 4 times, but would produce an eight-fold increase in body mass. This is very important in terms of metabolism, particularly in warm blooded animals, because the greater the area of body surface in relation to body mass, the greater the metabolic rate will have to be in order to replace the lost heat. Consequently, smaller animals have to burn more oxygen and produce more energy in order to maintain their body temperature: or put a different way, a larger body retains heat much more efficiently, and has to generate less heat per unit of body mass than a small body, thus explaining why larger animals burn energy more slowly.

The decrease in metabolism that accompanies increasing body size can be shown in other ways. For example, if we compare the respiratory requirements of a submerged water shrew with a whale, it is clear that the whale can remain under water holding its breath far longer than a shrew – proving the basic principle that small animals must respire at higher rates than large animals. A similar example is heart rate. For instance, the heart of an elephant beats with a ponderous thud approximately every two seconds (or 30 times per minute) compared to a mouse which has a resting heart rate of around 300 beats per minute. Thus, it is clear using these two simple examples that big animals use *relatively* less energy to keep alive compared to smaller animals.

One of the first biologists to show this relationship between size, reduced metabolism and increased longevity was Max Rubner, who in 1908 compared the rates of basal metabolism of humans, horses, cows, cats and guinea pigs. During the course of his calculations, however, Rubner also stumbled upon a more surprising relationship. He found that by the time these animals had reached the end of their lives, they had all used up roughly the same amount of energy. To be more precise, for each pound of body weight, Rubner showed that each animal used up between 30 and 55 million calories of energy. Only man was out of line, expending 365 million calories per pound, during an average life. These findings suggested to Rubner, that with the exception of man, animals appeared to have a certain amount of energy or 'living' allocated to their life span. They could use it quickly, or slowly, but the total amount of energy that could be expended by each unit mass of body tissue over life was essentially the same. This hypothesis which later become known as the rate of living theory, quickly become the subject of a great deal of research that has continued up to the present day.

Some of the first experimental evidence in support of Rubner's theory was published in 1917 by Jaques Leob and John Northrop who attempted to manipulate the life span of fruit flies by rearing them in different temperatures. Insects, along with many other kinds of animals including reptiles, amphibians and fish, are 'cold blooded' and consequently their pace of life tends to be heavily influenced by the external temperature. When placed in the cold, these animals become sluggish because the speed of their biochemical reactions slows down, and when they are warmed-up their metabolism increases. This does not occur with mammals and birds because their bodies have to be kept constantly warm: they have developed, in effect, a biological thermostat to ensure that their body temperature is carefully maintained. Thus, warm blooded animals have a fairly stable rate of metabolism which is not significantly altered by the external temperature. Indeed, if mammals and birds experience a drop in body temperature of only a few degrees they quickly lose consciousness and die. However, a similar drop in temperature in cold blooded animals will result in their metabolism slowing down. Cold blooded animals, therefore, make good models by which to test the effects of metabolism on life span. In short, if metabolism is linked to longevity, then slowing down metabolism by cooling the external temperature should increase life span. This is exactly what Loeb and Northrop set out to show by rearing flies in different temperatures.

The results confirmed the general prediction. For example, Loeb and Northrop found that when fruit flies were reared at 20°C they lived on average for 54 days; while those kept at a temperature of 25°C lived 39 days; and those maintained at 30°C lived only 21 days. Similar findings have been found in a wide range of cold blooded animals including roundworms, water fleas, cuttlefish, sea squirts and fishes. As a general rule, it appears that all cold blooded animals live longer at the lower end of the temperature range, with each increase of 10°C in external temperature resulting, on average, in a 50% decrease of life span. These results pointed to a clear conclusion. A basic mechanism controlling length of life appeared to be the rate at which chemical reactions occurred in the tissues and cells of the body: the faster the metabolism the shorter the life span.

In 1928, Raymond Pearl of John Hopkins University, introduced the term "rate of living" to describe this relationship between aging and metabolism. Pearl extended the work of Loeb and Northrop and showed that the effect of increased temperatures on fruit flies not only raised their rate of metabolism, but also speeded up growth and development. For example, at warmer temperatures, eggs developed faster; juveniles grew more quickly; and adult flies matured, reproduced, senesced and died earlier. Put another way, two groups of flies reared at different temperatures were physiologically different in age despite being the same age chronologically. Pearl's explanation was similar to that of Rubner's in that temperature increased the "rate of living"; and that a high rate of living shortened the life span because the animal's store of "life force" was used up more quickly. Pearl concluded that all species are born with the capacity to expend a finite amount of energy: they can either use it up quickly and age faster, or spend it more slowly and live longer.

In support of his theory, Pearl even demonstrated that the tiny water flea known as *Daphnia* had a specific allocation of heart beats by which to exist. Pearl found that there was a sex difference in heart rate in *Daphnia* with males producing an average of 4.3 beats per minute and females producing 3.7 beats. Intriguingly, the male had an average life span of 37.8 days compared to the female which was longer at 43.8 days. When Pearl multiplied the heart rate by the length of life he found that the actual number of heart beats produced by males and females was almost identical. Furthermore, when Pearl raised the temperature of *Daphnia* from 8°C to 28°C, he found that this increased the heartbeat fourfold and decreased the life span by 80%. This appeared to be a perfect example of rate of living.

PROBLEMS WITH THE RATE OF LIVING THEORY

As with most theories, however, things are not as straightforward as they first seem. In particular, the work of John Maynard-Smith at University College London, in the early 1960s, has cast considerable doubt on the rate of living theory. In his experiments, fruit flies were kept at different temperatures for part of their life, and then switched to a new temperature for the rest of their life span. Maynard-Smith found that when fruit flies were maintained at 20°C for the first half of their life, and then transferred to a temperature of 26°C, they actually lived the same length of time as flies that had been kept continuously at 26°C. This contradicted Pearl's rate of living theory which would have predicted that the transferred flies should have lived longer than those kept continuously at the higher temperature (although not as long as those kept continuously at the lower temperature). In addition, Maynard-Smith found that flies kept at 26°C and then transferred to a temperature of 20°C, had life spans that were similar to flies that had been kept throughout their lives at 20°C, again contradicting Pearl's theory.

To explain these results, Maynard-Smith proposed that a fruit fly's life span could be divided into two stages: an aging phase that is unaffected by changes in external temperature (or changes in metabolism), and a dying phase, which in contrast to the earlier stage was sensitive to the manipulation of temperature. At first sight, these findings appeared to show that the rate of living theory only applied to the second half (or dying phase) of life. However, other explanations were just as plausible. Indeed, rather than accounting for these findings in terms of metabolism, Maynard-Smith preferred to use the concept of "vitality". He suggested that a younger animal is much better able to cope with physiological stresses such as high temperature because it has greater inherent vitality. However, as the animal ages it loses this inner strength, and its resistance to stress (which includes adverse temperature conditions) declines. Put simply, fruit flies (and presumably other animals) are less able to cope with temperature stress during the second half of their life, and consequently such stress will have a much greater bearing on mortality and length of life.

A similar type of experiment was also undertaken by Roy Walford at the University of California who examined the effects of rearing the Argentine pearl fish *Cynolebias bellotti* at different temperatures. Most fish have relatively long life spans which limits their usefulness in aging research, but the pearl fish only lives for about a year and a half. In its natural habitat, *Cynolebias bellotti* is found in small ponds that during the summer dry up leaving the fish to die. Before this happens, however, the female lays her eggs in the soft mud at the bottom of the pond, and these remain dormant until the rainy season, when the pond fills up again and the eggs are hatched. When Walford kept these fish in the laboratory under conditions that closely matched their natural surroundings, he found that they still continued to live their normal life span of just over a year. Moreover, they exhibited clear signs of aging including a humped back and a decrease in their length which appeared to be due to arthritic degeneration of the spine.

But these age-related changes could be modified by changing the temperature of the water. For example, Walford found that when his fish were kept at 20°C they had average life spans of 14 months, whereas fish kept at 15°C had life spans of 19 months. In other words, these results were in accordance with the rate of living theory. However, not all of Walford's findings were so easily explained. For example, in contrast to the predictions of the rate of living theory, the fish reared at 15°C actually grew faster and ended up bigger than the fish raised at 20°C, suggesting that the colder water had not exerted its effect by simply slowing down metabolism. Even more problematical for the rate of living theory was the finding that fish kept for 8 months at 20°C and then switched to a temperature of 15°C had an average life span of 23 months; 4 months longer than for fishes continually kept at 15°C. In fact, the longest-lived individual fish of this group lived for nearly 40 months which was double that of the longest-lived fish under normal conditions. Again, it is hard to reconcile these findings with a metabolic explanation, and instead Walford suggested that the answer might lie with the immune system. The immunological theory of aging is covered in chapter 10, but essentially it explains aging in terms of a gradual immune reaction that takes place against the tissues of the body. In line with this idea, Walford has pointed out that fish kept at lowered temperatures show a depressed immune response, which in his view might be expected to reduce autoimmunity and thereby slow down aging.

THE RATE OF LIVING IN MAMMALS

It is not as easy to test the rate of living theory in warm-blooded animals because they are able to maintain their body temperature at a constant level independent of external fluctuations. Indeed, changing the external temperature may alter the superficial skin temperature of warm-blooded animals, but their core body temperature remains relatively unaffected. In fact, raising temperature will tend to lower the metabolic rate which, of course, is opposite to what happens in cold-blooded animals. Furthermore, it has been shown that rats subjected to a low environmental temperature (9°C) throughout their lives, compared to those maintained at a high temperature (28°C), have a shorter life span and an impaired ability to regulate body temperature at the end of their lives. In other words, it is meaningless to subject warm-blooded creatures to the same type of experimental conditions used to modify metabolism in cold-blooded animals.

Consequently, most attempts to understand the relationship between metabolism and life span in warm-blooded animals have tended to be more theoretical, and concerned with exploring Rubner's original hypothesis linking increased body size with reduced metabolism. And, in general, the theory has confronted many difficulties. For example, one problem with Rubner's work is that his calculations were based on the life spans of only five domestic animals (cat, dog, pig, cow and horse). It soon become clear after Rubner published his theory, however, that there were inconsistencies when other types of animal were examined. For example, Rubner's theory would predict the gorilla to be the longest lived primate, yet man lives, on average, twice as long as the gorilla which rarely lives beyond 35 years. Moreover, the chimpanzee which is significantly smaller than the gorilla lives for roughly the same amount of time despite having a similar level of metabolism. In other words, Rubner's theory does not work very well for primates.

Other types of mammal also show discrepant results. For example, it is known that marsupials have basal metabolic rates that are approximately 70% that of equivalent sized mammals, yet they are shorter-lived which is opposite to that predicted by the rate of living theory. Perhaps even more remarkable are the life spans of bats, which are a mammalian species living on average four times longer that their rates of metabolism would predict. Birds are yet another type of warm blooded animal that do not fit neatly into Rubner's theory. For example, birds generally have a higher basal metabolism than mammals which may be due to their somewhat higher (approx. 3°C) body temperature. Thus weight for weight, birds should have shorter life spans than mammals, yet the reverse tends to be true and a number of birds have very impressive life spans. For example, the golden eagle may live 46 years, a parrot 50 years, a starling 15 years, and a wren 7 years. This is despite the fact that birds typically use up 4 times as much energy as similar sized mammals during their life time.

It is clear, therefore, from the preceding account that there is no simple relationship between longevity and metabolic rate. Indeed, species with low or high metabolic rates appear to have developed short or long life spans depending upon how they have evolved. And, as we have seen in chapter 2, it is clear that there is more to the evolution of life span than metabolism. In fact, animals have evolved many physiological and cellular mechanisms that protect against the effects of metabolism. For example, increased metabolism will increase the number of chemical and free radical reactions, the accumulation of waste, and the amount of cross-linking between molecules. However, to combat these damaging processes, mammals have also evolved protective mechanisms such as DNA repair, and antioxidant enzymes including superoxide dismutase and catalase. Thus, whilst metabolism is undoubtedly linked to aging, it is clear that the relationship is not as simple as Rubner first believed. In short, there are many other variables to take into account.

It is interesting to point out that a similar conclusion was reached by the work of George Sacher (see also chapter 2) who found that four variables appeared to account for 85% of the life span variation in animals. In order of importance these are brain weight, metabolic rate, body weight and body temperature. Thus, in Sacher's view metabolism by itself is not necessarily a good predictor of a species life span although it undoubtedly has an important influence. But again, we must be cautious about accepting these findings without reservation. For example, bats do not have especially large brain sizes in relation to their bodies, yet as we have seen, they have unusually long life spans. Thus, it seems that there is an exception to every rule.

THE RELATIONSHIP BETWEEN METABOLISM AND EXERCISE

As we have seen, it is difficult to use mammals to test the rate of living hypothesis because they are warm blooded, and it is not possible to influence their metabolism in a meaningful way by altering their external temperature. But, in theory, there might be one way of changing metabolism – and that is by exercise. Indeed, metabolism increases dramatically with exercise, largely as a result of changes in muscle tone. For example, walking slowly at 2.6 miles per hour is sufficient to double our basal metabolic rate; sexual intercourse can increase it three-fold, and jogging can increase energy usage six times above normal. In fact, during hard physical exertion our metabolism and oxygen intake may increase some fourteen times above its basal rate. Thus, exercise would appear to be at least one way we can examine the effects of metabolism on life span in warm-blooded animals. In short, if the rate of living theory is correct, then we might expect exercise to hasten the aging process.

However, a number of studies have looked at the effects of exercise on the longevity of laboratory animals and, in general, they have found that providing the exercise regime is started early in life, the effects are likely to be beneficial rather than negative. For example, in one study where both male and female rats were forced to run 840 meters in 10 minutes every day, the exercised rats lived some 200 days more than controls, despite the fact that they had a 25% increase in resting metabolism. This increase in longevity appeared to be largely due to a reduced vulnerability to disease. In another study, rats were forced to run for 20 minutes each day, 5 days a week, with the exercise regime being initiated at one of four ages ranging from 120 to 600 days. The results showed that the groups started on the exercise at 120 and 300 days of age lived longer than controls, whereas the groups started at 450 and 600 days lived less, suggesting that there was a 'threshold age' beyond which it was not advantageous to commence exercise.

Similar types of experimental studies with humans are not feasible, and consequently researchers have had to resort to alternative strategies to see whether there is any relationship between exercise and longevity. One way of addressing the issue is to look at the life span of athletes, based on the reasonable assumption that they will have undertaken more life time exercise than non-athletes. A favourite type of subject for this type of study has been college oarsmen; partly because of the ease of collecting reliable data from universities to which they belonged, and since rowing is considered to be a particularly strenuous form of exercise. Indeed, a number of studies published around the turn of the century showed that college oarsmen, on average, enjoyed an increased life span of around 2 to 3 years. A more recent study published in 1972, which looked at the longevity of Harvard and Yale oarsmen from 1882 to 1902, found even more striking results, with oarsmen obtaining an average life span of 67.9 years, compared to a random sample of their classmates that lived 61.6 years. But not all studies have found that exercise confers greater longevity. For example, a study that looked at Danish athletic champions from 1880 to 1910 showed that although there was a significant lower mortality for the athletes under the age of 50, there was no difference in mortality after that age. In other words, increased athletic activity early in life did not appear to extend life span later on.

One problem with this type of research is that athletic activity in young adulthood does not necessarily equate with life time exercise since it is clear that many sports men and women stop regularly exercising as they get older. One sport that may provide an exception to this rule is cross-country skiing. Not only is this a high endurance sport requiring very high levels of oxygen intake, but more importantly, the active career of a skier typically extends over 20 years, and a large proportion of former champions retain skiing as a hobby. A study by M. Karvonen in 1956 which examined the mortality of 396 Finish champion cross-country skiers found that they gained some 6 to 7 years of longevity compared to normal skiers. In addition, they tended to have low blood pressure and were much more physically active in old age. Thus, in this particular case, exercise appears to have a beneficial effect on longevity.

Indeed, contrary to what the rate of living hypothesis predicts, common sense tells us that exercise is good for us. For example, people with highly active occupations such as lumberjacks, construction workers and farmers do not appear to age any faster than normal. Similarly, many centenarians have led physically active lives and have often worked well into old age. This has also been confirmed by the work of Dr Alexander Leaf who has found three remote communities in the world (Abkhazia in the Caucasus Mountains, Hunza in Northwestern Pakistan, and the village of Vilcabamba in the Andes Mountains) where people live to an unusually old age. These communities share a number of characteristics including a very high level of physical activity which involves active farming and tilling of the ground well into extreme old age. Thus, it is difficult to view physical activity as anything but beneficial to the individual. Indeed, this has also been confirmed in a number of other studies including a large scale study undertaken in the USA which assessed the level of physical activity in 1,909 men and 2,311 women who were aged 35 to 69. These subjects were then followed for a 14 year period, during which time it was found that there was a direct link between lack of exercise and mortality from heart disease. It is clear, therefore, that on an individual basis, decreasing one's rate of living by cutting down on exercise is much more likely to be associated with health risks than with any benefit.

Summary

- The metabolic rate (the sum of all chemical reactions taking place in the body) can be worked out by measuring the amount of energy that is lost by the body as heat, or be measuring the body's consumption of oxygen.
- The amount of energy used under resting conditions is normally called the basal metabolic rate (BMR).
- On average the human body reduces is basal energy expenditure by about 12 calories every year after the age of 30. This is probably due to changes in the body's composition that occur with aging.
- Although there does not appear to be a decline in the basal metabolic rate with aging, there is a decline in the body's efficiency to use energy over and above basal levels.
- The sites of energy production in the cell are the mitochondria. Research involving insects shows that degeneration of mitochondria is the main cause of flight muscle decline. However, in mammals the role of mitochondria in aging is less clear since mitochondria are normally able to replicate themselves.
- Larger animals tend to have a slower metabolic rate than smaller animals because they are generally much better able to conserve heat.
- In 1908, Max Rubner calculated that all animals appeared to use up the same amount of energy per unit of body weight during life. In

1928, Raymond Pearl used the term "rate of living" to describe this relationship.

- Research looking at the effects of rearing fruit flies in cold environments (which reduces their metabolic rate) originally appeared to support the rate of living theory, although later work by Maynard-Smith cast doubt on these findings.

- There are now known to be many problems with Rubner's theory especially with regard to warm blooded animals. For example, bats live on average four times longer than their size or metabolism would predict, and many primates (including man) do not conform to its expectations.

- George Sacher has shown that 4 variables appear to account for 85% of the life span in animals and in order of importance these are: brain weight, metabolic rate, body weight and body temperature.

THE HORMONAL BASIS OF AGING

You are only as old as your glands

–Dr. Brinkley

Aging as we see it in the total animal, may be more a function of the breakdown in integrative mechanisms than of changes in individual cells, tissues, or organs.

–**Nathan Shock**

Animals co-ordinate and control the activity of their body organs through two major systems of internal communication – the nervous system and the endocrine system. The first of these systems conveys high-speed signals along specialised cells called neurons whose function is to receive, transmit and interpret information (mainly by the brain) and then to initiate appropriate responses such as voluntary movement or control of vital body structures. In contrast, the hormonal system provides a much slower means of communication through the release of chemical messages carried in the blood or bodily fluids. The result is that some hormones produce their effects in seconds, whilst others may take hours and days. Throughout the animal kingdom, hormones are absolutely crucial in the control of growth, development, reproduction and homeostasis – all functions which as we have already seen have been linked in various ways to aging. In fact, if we examine the actions of hormones more closely, it can be seen that they exert their effects by regulating the metabolic functions of the body, including the rate at which chemical reactions take place in the cells themselves. It is therefore not surprising that ever since hormones were first discovered, they have been implicated in aging. Not only are they able to influence the biochemistry of practically every cell in the body (and in some cases the functioning of the genes), but it is also well established that levels of many hormones change with age, which has helped foster the widespread belief that hormonal replacement may bestow anti-aging or rejuvenatory benefits. Hormones must therefore be regarded as a prime suspect in our quest to understand the causes of aging.

THE ENDOCRINE SYSTEM

The endocrine system consists of a number of glands scattered throughout the body that secrete chemical substances called hormones (from the Greek *hormon* meaning to 'excite') which are released into the blood where they are transported to various parts of the body. More than 50 different hormones may be circulating through the blood at any one time, and each has its own special function. In vertebrates more than a dozen tissues and organs secrete these chemical messengers, the most important of which are the *thyroid* and *parathyroid glands* situated in the neck; the *thymus* placed in the upper chest; the *adrenal glands* located just above the kidneys; and the *gonads* including the testes and the ovaries. In addition, it is now known that hormones can also be released by organs such as the stomach, intestine and liver although traditionally these structures have not been regarded as part of the endocrine system. Hormones generally exert their effects by speeding up chemical reactions that are already taking place in the cells and, because they work in practically all tissues of the body, they have a large and diverse set of functions. Indeed, it can be said that there are few functions in the body that are not in some way influenced by hormonal regulation. In short, they are vital for life.

Although the endocrine glands are distributed throughout the body, most come under the control of a single master gland that is attached to the underside of the brain, and found just above the roof of the mouth. This all-important pea-sized structure is called the *pituitary gland,* and it releases a number of hormones whose role it is to regulate the other endocrine glands of the body. The pituitary gland actually consists of two separate glands; the anterior and posterior, which are structurally very different. The anterior pituitary is known to release at least eight hormones, most of which are called *trophic* hormones, and whose primary function is to control the release of hormones from other glands in the body. For example, the anterior pituitary secretes thyroid stimulating hormone which induces the release of thyroxine from the thyroid gland; and adrencorticotrophic hormone (ACTH) that acts to release cortisol from the adrenal cortex. It is largely on account of these trophic hormones that the pituitary is often called the master gland, although it also secretes two non-trophic hormones; growth hormone and prolactin, that have a direct effect (i.e. without the intervention of other endocrine glands) on the cells and tissues of the body. The posterior pituitary is somewhat different and only releases two hormones, vasopressin (involved in regulating blood pressure) and oxytocin (involved in uterine contraction and lactation). These two hormones will be not be discussed any further since neither have been strongly implicated in aging.

The pituitary gland is essentially the command centre of the endocrine system, and one of its most important functions is to maintain appropriate levels of its hormones circulating throughout the body. In order to do this, the pituitary must constantly monitor hormone levels and respond by increasing their output if levels drop (or by decreasing secretion if their levels increase). In this way the pituitary gland is able to control the hormonal environment of the body. However, there is a problem with such a relatively simple system because it is limited in the way it can adapt to new contingencies or biological demands. For example, during times of high stress the body adapts by increasing cortisol secretion to meet the new metabolic demands of the body. But how does the pituitary gland know when stress is occurring? Similarly, all hormones are released in their own distinctive circadian pattern. But how does the pituitary gland know what time it is? The answer to both questions is that the pituitary receives input from a brain structure called the hypothalamus which has access to higher order information. In fact, the pituitary gland and hypothalamus are joined together by a stalk of tissue that contains its own blood system, and into which the hypothalamus secretes its own chemical messengers (called releasing or inhibiting factors) that modulate the pituitary's release of hormones. The pituitary gland may be the master gland, but it is the hypothalamus that exerts overall executive control when important changes need to be made to the body.

The combination of the hypothalamus and pituitary gland working together means that the control exerted over hormone secretion is complex and finely tuned. This can be illustrated with the following example involving thyroid stimulating hormone (TSH) that is released when the body needs to increase its metabolism or temperature. As core body temperature starts to drop, body temperature detectors in the body start sending messages to the brain telling it to increase heat production. In response, the hypothalamus secretes a releasing factor that acts on the pituitary to release TSH into the bloodstream. When TSH reaches the thyroid gland, thyroxine is released causing an increase in metabolic activity and a corresponding increase in heat production. At this point, the body detectors start signalling increased body temperature messages to the brain, and the increased levels of thyroxine will feed back to the pituitary (and the hypothalamus) which acts to suppress further release of TSH. This example shows that hormonal secretion is often the end result of both hormonal and neural feedback that is integrated in the hypothalamus and the pituitary gland.

The fact that hormonal control is under the influence of the pituitary gland and hypothalamus, has led some researchers to point out that these structures are situated in an ideal position to act as biological clocks that might conceivably allow them to time the decline of hormonal function. The beauty of this idea is that it offers a simple explanation of how aging could be programed into an organism, since the program would only need to be encoded into the pituitary gland and hypothalamus, and not in every cell of the body. Although there is not a great deal of evidence to support this theory there is, nevertheless, still a suspicion that because of their pivotal position in the functioning of the body, the pituitary gland and hypothalamus still hold many vital clues about the secrets of aging.

THE EFFECTS OF PITUITARY GLAND REMOVAL ON AGING

If aging has a hormonal basis then it should be possible to manipulate its rate by changing the levels of hormones in the body. Perhaps the most simple and dramatic way of doing this is to remove the master gland itself. The removal of the pituitary gland (otherwise known as a hypophysectomy) is relatively simple to undertake in laboratory rodents, and it results in a severe decrease of hormone levels, particularly those secreted from the thyroid, adrenal cortex and gonads. Surprisingly, hypophysectomy produces few harmful effects, and providing the animal is given regular cortisone injections (without which it would quickly die), it typically survives the operation in good health. Perhaps the most remarkable finding of all, however, is that the operation produces animals that live longer than usual. Much of this research has been undertaken by Arthur Everitt working at the University of Sydney in Australia, who has shown that removal of the pituitary gland increases longevity by about 15%. For example, in one study where rats were hypophysectomised at 60 days, their average life duration was 916 days compared to 785 days for non-operated animals. Everitt has kept records from work that has spanned over 25 years, and has found that the longest-lived lesioned rat of all reached the age of 1,352 days, compared to 1,201 days for the longest-surviving normal animal. In short, this and other evidence suggests that hypophysectomy increases both the average duration of life and its maximum duration. However, this operation has to be performed at an early age since removal of the pituitary gland in later life (at 400 days) has been shown to have a life shortening effect.

How then does removal of the pituitary gland produce these effects? One of clearest benefits of the operation is on the suppression of tumour formation. Laboratory animals such as rats and mice usually have a high mortality from tumours – particularly those of the endocrine glands including thyroid, testes and ovary. This cause of death, however, is not a characteristic feature of pituitary-lesioned animals. For example, in one study that looked at pathology at the time of death, tumours were found in 67% of unoperated male rats, and only 12% of those that had been hypophysectomised. It might be expected from this finding that removal of the pituitary gland simply slows down the development of disease, but it appears that there is much more to hypophysectomy than this. For example, removal of the pituitary gland results in significantly slower growth, a marked decrease in metabolism, and a slowing down of cardiovascular function. On this basis alone, pituitary-lesioned animals might be expected to show a slower rate of aging, and indeed, there is considerable evidence to justify this claim. For example, it has been shown that collagen (see page 125) ages twice as slowly in hypophysec-tomised rats; and this reduced rate of aging is further supported by the finding that the age-related loss of eggs from the ovary in female rats is halved in those subjected to pituitary removal. In addition, these ani-mals show other signs of reduced aging including improved kidney func-tion, decreased aorta wall thickness, improved wound healing and increased thymus size. All of this evidence strongly supports the idea that pituitary gland excision slows down the process of aging.

However, these impressive results need to be viewed with a certain degree of caution since there are also major drawbacks to the operation which makes it far from ideal as a form of life extension. In particular, animals with lesions of the pituitary gland are poor at tolerating stress and coping with low temperatures, both of which can quickly lead to death. Because of this, hypophysectomised animals have to be looked after carefully in stable warm conditions and subjected to little dis-tress. Consequently, in anything but ideal laboratory conditions, the beneficial effects of pituitary gland removal are not observed.

What then do the results of these studies tell us about hormones and aging? Everitt believes that the secretion of pituitary hormones greatly speeds up the aging of their target tissues. In other words, hormones hasten the aging process and pituitary removal slows down the speed of the decline. At first this notion may sound implausible, but there is undoubtedly some degree of truth in the theory as hormones do clearly speed up the metabolic processes of the cells they interact with and, as we saw in the previous chapter, increased metabolism is linked with aging. Despite this, we must view this conclusion with a certain amount of caution because hormones are clearly needed for the proper functioning of the body under normal conditions. Furthermore, while it is not ethically possible to produce hypohysectomy in humans, in the few cases where humans have suffered pituitary gland damage there has often been accelerated and premature aging. In particular, a very rare disorder known as Glinski-Simmond's syndrome which is caused by destruction of the anterior pituitary is known to result in an aged appearance with wasting and greying hair, accompanied by early senility. Thus, the findings with rats may not necessarily apply to humans. It is also possible that hypophysectomy slows down aging in ways unrelated to decreased hormone levels. For example, hypophysectomised animals eat significantly less than normal and, as we shall see in chapter 11, reduced weight and food restriction is also a powerful determinant of increased longevity.

The Thyroid Gland and Aging

If the pituitary gland is involved in aging then the question arises as to what endocrine gland it is having its main effect on. One possible candidate is the thyroid gland which is a smallish 'H' shaped tissue mass situated just below the Adam's apple. The thyroid gland secretes two main hormones, thyroxine and triiodothyronine, both of which are continuously at work controlling the rate of metabolism in our bodies. Put simply, these hormones work by regulating the rate of oxygen consumption; which is another way of saying they control the speed of chemical activity in our body cells. Since aging has been linked to metabolism (see previous chapter) this gives us reasonable cause for suspecting an involvement of the thyroid gland in aging.

Thyroid hormones are at work throughout life. During childhood they are particularly important for growth, and too little thyroid stimulation can slow physical growth and delay the maturation of the brain, resulting in a condition known as cretinism. In contrast, increased release of thyroid hormone may make children grow much faster than normal. In adulthood, the main function of the thyroid hormones are to control the body's metabolism and heat production; and they do this by increasing the metabolic activities of practically every tissue in the body with the exception of the brain. Although other hormones such as growth hormone and testosterone can also influence metabolism, the thyroid hormones are by far the most important. For example, secretions of thyroxine may increase the metabolic rate by 100%, and decreased secretion may cause the metabolic rate to drop by as much as 40%. By increasing the rate at which cells can use oxygen, thyroid hormones determine the rate at which food is broken down to produce energy and heat. Thyroid hormones also increase protein synthesis; in addition they increase the reactivity of the nervous system – resulting in increased blood flow, a strong heartbeat, and increased mental excitability.

In support of a thyroid involvement in aging is evidence showing that deficiency of thyroid hormones in adults produces a pattern of change that resembles aging in a number of ways. For example, an underactive thyroid gland produces a condition called myxoedema, characterised by thickened pale and wrinkly skin, loss of hair and increased greying. The limbs ache and the person feels run down and sluggish. The voice becomes gruff, hearing becomes dulled, and there is a tendency to increase weight. The person is also likely to feel cold even when placed in a heated room where others are feeling comfortable. Thyroid deficiency may also result in a reduced heart rate with a long term risk of angina pectoris due to narrowing of the arteries. Finally, there may be a tendency for mental slowing with impaired memory and slowing of speech. In short, these effects have a striking resemblance to many of the changes that accompany old age.

Because of this, it has been suggested that normal aging may be the result of declining or inadequate levels of thyroid hormones. Despite the obvious attractiveness of this theory, there is little evidence to support it. The biggest problem is that there is no reliable decline in blood levels of either thyroxine or triiodothyronine in healthy aged individuals. For example, in one study of several hundred healthy old people over the age of 60 years, no decrease of thyroxine was found, although 20% of individuals showed a moderate decrease of triiodothyronine. Since thyroxine is secreted in far greater amounts than triiodothyronine, this is reasonable evidence that thyroid levels do not significantly change with age. A second way of assessing thyroid function is by measuring the release of thyroid stimulating hormone (TSH) from the pituitary gland in response to injections of thyrotropin releasing factor (TRH) which is normally secreted by the hypothalamus (and controls TSH release from the pituitary). Several studies have shown that the amount of TSH produced in response to TRH is diminished in older men but not in women. While this finding shows an age-related decline in pituitary gland sensitivity in men, the fact that this change does not occur in both sexes shows that this explanation cannot be used as a general theory of aging.

It is interesting to note that around the turn of the century attempts were made to reverse the effects of old age by administering thyroid hormones. There is little evidence that this treatment worked and in some cases it even lead to fatalities. Administration of thyroid hormones increases the metabolic rate and makes the body use up its reserves of stored energy more quickly, resulting in the person getting thinner despite eating well. For this reason these hormones were once used as slimming aids (and were once even freely available over the chemist's counter) although this practice is rare today. One drawback of using thyroid hormones is that the heat generated by the increased burning up of oxygen makes the skin pink, warm and sweaty, and consequently heat intolerance may occur. Several other side effects including hyper-excitability and restlessness were also common, although some people apparently enjoyed this effect, and addiction to thyroid preparations was not unknown.

DONALD DENCKLA'S 'DEATH HORMONE'

Despite all the evidence presented above, uncertainty regarding the role of the thyroid gland in aging has never quite gone away. In particular, the suspicion exists that it is not so much the levels of thyroid hormones that are important, but rather that our cells become less sensitive to the effects of thyroid hormones acting upon them. A variation of this theory has been proposed by Donald Denckla who has argued that thyroid hormones gradually lose their effectiveness because the pituitary gland also releases a substance that, with aging, blocks their action in the cell. In other words, Denckla believes that the pituitary gland secretes a chemical that interferes with the action of thyroid hormones, which in turn causes aging. Not surprisingly, in some quarters this substance has been dubbed a 'death hormone'. If Denckla is correct, this would, of course, provide an alternative explanation as to why animals without a pituitary gland manage to live longer than normal (i.e., they produce no 'death' or 'aging' hormone).

As we have seen in the previous chapter, the question of whether basal metabolism declines with aging is a difficult one. If basal metabolism does decline then it must do so at a slow rate, and evidence showing normal levels of thyroid hormones with aging also supports this idea. However, Denckla decided to examine basal metabolism in a completely different way – by measuring the amount of oxygen consumed by animals that were anaesthetised. Because this is the lowest possible energy state that the animal can survive in, Denckla called the technique minimal oxygen consumption (MOC). In this state, oxygen consumption was found to be some 20–30% lower than the basal metabolic rate obtained when the animal was conscious and resting. But the most striking finding was the manner in which minimal oxygen consumption was shown to decrease in older animals. In short, this measure of metabolism fell by a remarkable 75% over the life span. In other words, whilst there may be some degree of doubt over the question of whether basal metabolism declines with aging, there is little doubt that minimal oxygen consumption falls dramatically over the same time span.

Denckla then set about trying to determine what endocrine structures were involved in regulating oxygen consumption in his anaesthetised animals. To do this he measured MOC in more than 2,500 rats of differing ages, and with various endocrine organs removed (adrenals, pancreases, ovaries, etc.). The results showed that only two endocrine structures were involved in controlling the amount of oxygen used by anaesthetised animals; namely the thyroid gland and the pituitary gland. Furthermore, the thyroid appeared to have the more important influence. Thus, in apparent contrast to the findings with basal metabolism, thyroid hormones were apparently controlling metabolism after all.

But there was still an unexplained problem. Why was there a decline in minimal oxygen consumption with aging when there were normal concentrations of thyroid hormones circulating in the body? Part of the answer appeared to lie with the sensitivity of the body cells, since Denckla showed that young animals were 3 times more responsive to thyroxine administration than adult rats. But why did this decline in sensitivity occur? The obvious answer was that the aging cells were becoming less efficient in their utilisation of thyroid hormones. However, Denckla ruled out this possibility when he found, to his surprise, that he could decrease the responsiveness of his anaesthetised animals to thyroxine by injecting them with extracts taken from the pituitary gland. This made little sense unless one hypothesised that some substance in the pituitary extract was blocking the action of thyroxine. Denckla called this imaginary substance declining consumption of oxygen factor (DECO) and proposed that this substance was responsible for slowing down metabolism with aging.

Denckla's work remains highly controversial, and there appears to be a considerable reluctance on behalf of the scientific community to accept the existence of DECO. Indeed, this substance has never been isolated, and Denckla has not published any research for nearly 20 years. The reluctance of other scientists to accept the DECO theory is, perhaps, not too surprising since it would dramatically change the way most scientists think about aging (and it would be at odds with most of the theories discussed in this book). Indeed, the concept of a death hormone remains incredible to many, and as an explanation of aging it seems too simple to be true. Nevertheless, many of Denckla's findings are unexplained, and the important question of why the pituitary gland is apparently able to block the effects of thyroid stimulation still remains unresolved. Despite this, few scientists appear to be interested in pursuing these questions further.

THE ADRENAL GLANDS

Another endocrine structure that has been implicated in aging are the adrenal glands, situated just above the kidneys. These are composed of two separate and independent glands: an inner core called the medulla and an outer layer called the cortex, both of which produce different hormones. The adrenal medulla is more like a knot of nervous tissue than a gland, and releases adrenaline (epinephrine) directly under the control of the autonomic nervous system. Adrenaline tends to be released under emergency or sudden periods of stress and acts to accelerate heart rate, raise blood pressure, dilate the air passages of the lungs, increase mental alertness, and to make more blood sugar available for the body tissues and organs. In short, the adrenal medulla prepares the body for physical action, particularly in response to a threatening or stressful situation.

The adrenal cortex is a more typical hormone gland and is governed by chemical rather than neural control. The cells of the adrenal cortex produce around two dozen types of hormones (collectively known as corticosteroids) that are steroids and synthesised from cholesterol. The most important group of these, for our purposes, are the glucocorticoids that include cortisol, cortisone and corticosterone. Unlike the adrenal medulla, the release of these hormones, is controlled by the pituitary gland via its release of adrencorticotrophic hormone (ACTH). Glucocorticoids are absolutely essential for maintaining the correct chemical balance of the body in the face of change and adversity, and without their influence we would quickly die. For example, they help to keep blood glucose levels constant; and they also control the balance of salt throughout the body thereby helping to maintain blood pressure and cellular volume. They also play a role in increasing metabolism, and in promoting a wide range of adaptive responses, including the breakdown of protein for energy when the person is suffering periods of prolonged stress.

The importance of the adrenal glands can be seen by examining the medical conditions in which they become dysfunctional. For example,

Addison's disease results from damage to the adrenal glands (often as a result of autoimmune disease) that leads to cessation of glucocorticoid release. Although it is extremely rare, two famous people who have suffered from the disease are Jane Austen and John F. Kennedy. Today it is treatable with injections of glucocorticoids; but without medical intervention the disease leads to muscle wasting, poor circulation, hypotension, loss of blood with dehydration, and a discolouring of the skin. However, even with glucocorticoid treatment there is a tendency for these people when faced with a severe stressful event to fall into an 'Addison's crisis' with a sudden and potentially fatal drop of blood pressure.

The opposite of Addison's disease is Cushing's syndrome, a disorder often produced by tumours of the pituitary gland or adrenal cortex, resulting in an increased release of glucocorticoids. Amongst its many symptoms are increased blood glucose levels, severe muscle wasting, demineralisation of the bones leading to osteoporosis, and salt retention that facilitates the development of hypertension. Interestingly, these types of problem are not dissimilar to the ones that tend to develop in old age, and this has led some investigators to speculate that there may be an adrenal link with aging. This theory is in fact relatively easy to test because if it is true, then we would expect levels of glucocorticoids to increase with aging rather than show a decline.

GLUCOCORTICOIDS AND AGING

The situation regarding age-related changes in glucocorticoid levels is still far from clear despite many decades of work. Most studies have shown that basal levels of glucocorticoids, including cortisol, do not change with aging, and that old individuals have about the same amount of adrenal hormones circulating in their blood as young. However, these results are not as straightforward as they first seem because some studies have also found that the adrenal gland appears to *release* less cortisol in older individuals. This finding only makes sense if cortisol is existing for longer in the blood before being broken down and excreted. Indeed, this theory has been supported by work showing that cortisol may last up to 40% longer in elderly individuals compared to young. This would seem to imply that the control of adrenal cortex secretion is not changed with aging, but rather the rate at which the body cells use up glucocorticoids is slowed down.

However, the view that basal glucocorticoid levels are unchanged in old age is currently undergoing revision; recent work has suggested that glucocorticoid levels, particularly cortisol, may well be *increased* in very old individuals, especially those who are octogenarians or greater. Although this is not an inevitable feature of human aging, this finding accords with a lot of animal research that has consistently found higher levels of glucocorticoids in aged rats and primates. Although the reason for this increase is not totally clear, it is probably linked to feedback inhibition of the pituitary gland. As noted in the introductory part of this chapter, when glucocorticoids are released from the adrenal cortex they act on to the pituitary gland to inhibit further secretion of ACTH. In extreme old age, however, it seems that the pituitary gland becomes less sensitive to this hormonal feedback, and continues to secrete ACTH even when the blood levels of glucocorticoids are still quite high. Thus, the similarity between the symptoms of Cushing's syndrome and aging may not be totally unrelated after all.

AGING, STRESS AND THE ADRENAL RESPONSE

Results from animals studies have also revealed yet another important difference between young and old subjects when it comes to glucocorticoid secretion. Although both young and old alike secrete glucocorticoids in response to stressful events, they differ considerably in the speed with which they terminate this secretion. For example, in young animals the secretion of corticosterone is halted almost immediately with the removal of the stressor; in contrast, aged rats will typically secrete corticosterone for many hours after the stressful event has finished. This enhanced response appears to be largely due to the increased release of ACTH from the pituitary gland which may be increased fourfold in the stressed and aged rat. The relationship between stress and pituitary function is a difficult one to examine experimentally in humans – for obvious ethical reasons – but what little evidence there is which has examined stress in older people, suggests that they are also susceptible to similar increased ACTH responses.

The reasons for a marked increase of stress-related ACTH in old animals has been examined and discussed by Robert Sapolsky of Stanford University, in his excellent book *Why Zebras Don't Get Ulcers*. He has shown that the main cause of this effect lies far beyond the level of the pituitary gland, and begins with a higher brain structure called the hippocampus, whose function is generally more associated with a role in emotion and memory (e.g., it is known that damage to the hippocampus produces severe amnesia). It is clear, however, that the hippocampus exerts an important influence over the pituitary gland. In fact, the hippocampus is very sensitive to circulating levels of glucocorticoids and contains receptors for these hormones. Moreover, when it is stimulated by glucocorticoids the hippocampus then acts to turn off the pituitary gland's release of ACTH. Why the hippocampus is so intimately involved with the feedback inhibition of glucocorticoid secretion is still far from clear, although it may be linked to its mnemonic function in remembering previous stressful situations.

The involvement of the hippocampus in this process of pituitary control has, however, certain drawbacks – not least because the hippocampus appears to undergo considerable degeneration with aging. In fact, it has been reported that the hippocampus may lose up to 20% of its neurons by old age, which is a figure far greater than that for most other areas of the brain. Since the hippocampus normally acts to inhibit the pituitary gland's secretion of ACTH, it can then be seen that one consequence of its cell loss will be a reduced ability to inhibit the pituitary gland; and this may be one reason why ACTH levels become increased with stress and aging. But why is the hippocampus so susceptible to deterioration? The somewhat startling answer according to Sapolsky's work is because of its sensitivity to glucocorticoids!

The susceptibility of the hippocampus neurons to glucocorticoids was first reported in the 1960s when scientists found that laboratory animals injected with high amounts these hormones showed marked damage to the hippocampus. Further studies found that similar damage could also result from the secretion of the body's own glucocorticoids, particularly by the increased levels that typically followed a prolonged period of stress, with neuron loss being most marked in aged rats. In contrast, if laboratory rats had their adrenal glands removed early in life (and kept alive with low dose injections of glucocorticoids) hippocampal degeneration was significantly reduced. The alarming conclusion that follows from this research is that stress, particularly in later life, is likely to lead to degeneration of certain brain structures, and consequently speed-up the aging of the brain. Moreover, when the hippocampus is damaged, the animal secretes even more glucocorticoids (because its inhibition of the pituitary gland is reduced) which in turn is likely to damage the hippocampus further, and so on. In other words, a degenerative cascade occurs which can only multiply as aging progresses.

This work has a number of important implications. For example, the hippocampus is known to be particularly susceptible to damage from several sources including hypoglycaemia, oxygen deprivation, epilepsy, viral infections and ischemia (lack of blood flow) caused by cardiac arrest. Research has shown that even low levels of glucocorticoids makes such damage significantly worse. It appears, therefore, that it is increasingly difficult to protect our hippocampus as we age, and this might be one of the reasons why memory decline is such a serious problem in the elderly. The finding that people with Cushing's syndrome, (caused by an overactive adrenal gland) suffer serious memory problems is also consistent with this idea. It has long been known that stress is bad for health, but the realisation that it may also be causing brain damage, particularly in old age, adds a frightening new perspective to the issue.

Perhaps we should not be too surprised that glucocorticoids have been linked with neural degeneration. For example, animal studies have long shown that oversecretion of glucocorticoids can cause cardiovascular damage, but even more striking are the effects of glucocorticoids in causing death following semelparous reproduction. As we saw in chapter 4, there are some animals, including several species of fish and a few mammals (e.g., Australian marsupial mice) that are programed to die following reproduction. These animals typically show widespread tissue degeneration and organ atrophy within days of reproducing, and this has been shown to be due to a sudden and dramatic oversecretion of hormones from the adrenal cortex. Although nothing like this extent of glucocorticoid release occurs in humans, this example is nevertheless a grim reminder that glucocorticoids have the potential to do considerable harm if they are not regulated properly. And, of course, one of the problems with aging is exactly this type of homeostatic decline in many systems of the body (e.g., see pages 16 and 165).

DEHYDROEPIANDOSTERONE (*DHEA*)

An adrenal hormone that has attracted a great deal of interest in recent years is dehydroepiandosterone (DHEA). The function of this hormone has never been clearly established and this has been one reason why it has tended to be overlooked in the past. Remarkably, DHEA reaches a concentration in the blood of young adults that is 10 times that of any other steroid hormone produced by the adrenal cortex. And even more striking is the fact that between the ages of 35 and 70 years, the levels of DHEA decline so dramatically that by old age its levels may be 20–10% of that found in youth. In other words, DHEA is one of the few hormones that shows a significant decrease with advancing age, and it reaches its lowest levels at the end of life. Although its exact function is unknown, it is believed to play a role in the formation of the sex hormones testosterone and oestradiol (see next chapter), and in partial support of this theory, around 20% of DHEA in males is produced by the testes with the rest being made by adrenal cortex under the control of ACTH. Consequently, young males tend to produce more DHEA than women, although with aging they also show a much more marked decrease in its secretion.

During the 1970s a number of studies began to show that DHEA administration improved the conditions of laboratory animals that were genetically susceptible to obesity, diabetes and tumours (particularly those of the mammary tissues). This, in turn, led researchers to examine how DHEA could be exerting its beneficial effects. Although animals treated with DHEA tended to eat less than usual, it was found that they gained less weight per gram of food eaten, and moreover they demonstrated significantly improved glucose tolerance (see page 165). Thus, DHEA appeared to have a rejuvenating effect on certain biochemical and metabolic parameters of the body. Indeed, subsequent work has confirmed its impressive effects in animals, and it has been reported that high doses of DHEA reduce body fat by up to one third, prevent atherosclerosis, enhance the immune system, inhibit the development of certain autoimmune diseases, and extend the normal life span of mice by 20%. And all of this in addition to well documented effects on obesity, diabetes and tumour formation.

What about the effects of DHEA in humans? In 1986, a study published in the prestigious *New England Journal of Medicine* reported that low serum levels of DHEA was a significant risk factor for heart disease in older men; and that low levels were consistently found in men over 50 years of age who died early from any cause. Although this relationship did not hold for women, subsequent work nevertheless showed a link with breast cancer and low levels of DHEA in women who had passed their menopause. Thus, there seemed to be a clear DHEA link in both aging and disease. Further work has since shown reduced levels of DHEA in several illnesses including Alzheimer's disease. Furthermore, DHEA treatment has also been shown to protect against heart disease by reducing blood levels of cholesterol in young men. Thus, much of the human data seems to be in accordance with the animal findings.

Inevitably, these encouraging results have led to trials of DHEA in human aging. One of the first trials was undertaken in 1994, which looked at 17 women and 13 men between the ages of 40 and 70 years. All took 50 mg of DHEA by mouth at bedtime for a three month period. Although the treatment did not affect the amount of body fat or glucose metabolism, the subjects reported an increased sense of well-being, including greater energy, better ability to cope with stress, and improved quality of sleep. Positive responses were reported by 82% of women and 67% of men and, significantly, fewer than 10% of the volunteers reported these effects whilst taking a placebo (the study was designed so that the volunteers did not know what substance they were taking). These findings were also confirmed in a second study that looked at 8 men and 8 women aged 50 to 65 who were given 100 mg of DHEA for 6 months. Treatment at this higher dose was also found to increase lean body mass, and raise levels of certain biochemicals known to be involved in growth.

Not surprisingly, since the publication of these results, DHEA has attracted a great deal of media attention, particularly in the USA where it is freely available by mail order (in the UK it is only available on prescription). Moreover, several popular books have raised public awareness of this hormone with claims that DHEA prevents cancer and heart disease, restores sexual potency, improves memory loss, and halts middle-aged weight gain. In fact, it has to be admitted that it has more than just a popular appeal. It has been estimated that more than a quarter of medical practitioners participating in a recent conference held on DHEA, and whose findings were published by the New York Academy of Science, were actually taking DHEA for themselves! Yet, despite this, many other doctors have cautioned against its widespread use, arguing that much more research on its safety and long term effects needs to be undertaken. It remains to be seen who is right.

GROWTH HORMONE

Unlike most of the other hormones released by the pituitary gland, growth hormone acts on the tissues of the body directly without the intervening help of any target endocrine gland or organ. Moreover, the control of growth hormone release is particularly complex as it appears to be under the influence of several different feedback systems. Most of the time it is under the control of the hypothalamus which directs its release either by secreting a growth hormone releasing factor (called somatocrinin), or an inhibiting factor (called somatostatin). In other words, the hypothalamus has the dual role of acting both as a brake and an accelerator on growth hormone secretion. But other substances are also able to act on the hypothalamic-pituitary axis to influence the release of growth hormone, including the hormone melatonin (see next section), as well as blood sugar. While many factors can affect growth hormone release, the most important event of all appears to be sleep. A huge surge of growth hormone occurs during the first few hours of sleep, and this appears to be unrelated to any change in the hormonal balance of the blood.

Growth hormone, as its name suggests, has a vital role to play in early growth and development as it stimulates the growth of bones, muscles and various organs of the body. Indeed, if the body produces too much growth hormone during childhood, gigantism can occur, whereas too little may cause dwarfism. However, growth hormone also has many important functions in the adult. In particular, it promotes the uptake of amino acids into the cell and their incorporation into proteins (protein synthesis). It also encourages the breakdown of fat stores into fatty acids that can be used as energy. Thus, one function of growth hormone is the upkeep and restoration of body tissue, particularly proteins, most of which are relatively fragile and need to be replaced with use. This probably explains why growth hormone is released in such high amounts during sleep (e.g., this is the best time for restoration to occur) and why it tends to be released in response to vigorous exercise, particularly in young people.

GROWTH HORMONE AND AGING

In contrast to thyroid and adrenal hormones whose blood levels show little significant decline with aging, there is a marked age-related decline of growth hormone. The decline in secretion begins at around the age of 30 years, and on average (although it is highly variable), appears to fall by about 14% with each decade. The consequence of this decline is that by the time an individual reaches 80 years, the levels of growth hormone may have dropped by over 50% compared to the age of 30. This decrease in secretion affects primarily the night time surge with the result that aging tends to flatten out the circadian rhythm of growth hormone release. Two neurotransmitters used by the hypothalamus to transmit neural messages (noradrenaline and dopamine) are known to regulate growth hormone release, and levels of each are decreased significantly with aging in both animals and humans.

Furthermore, there appears to be a marked resemblance between many of the features of aging and changes that might be expected to come about through a decline in growth hormone secretion. For example, aging results in a reduction in protein synthesis, a decrease in muscle mass (with a resultant increase in body fat), and a loss of bone mass – exactly the types of biological activity that growth hormone is involved in regulating. The decline in lean body mass (or muscle) is particularly striking. On average the amount of lean body mass decreases by 27% in men and 15% in women from 25 to 70 years of age. This is accompanied by an increase in fat stores, particularly in the abdominal region, of approximately 18% in men and 12% in women. Therefore, there appears to be an excellent correlation between declining growth hormone levels and increasing fat tissue, and further evidence suggests that this relationship is more than incidental. For example, not only does decreased growth hormone appear to reduce muscle mass, but it seems that increased fat stores in men also produces a further reduction in growth hormone secretion (and a decrease in the amount of time it can exert its biological activity). Thus, there appears to be a reciprocal relationship between amount of fat and growth hormone; with increased body fat leading to greater growth hormone decline and less growth hormone contributing to the increased storage of fat. One can surmise, therefore, that reduced levels of growth hormone may well be one of the factors leading to increased fat storage with aging.

We might, therefore, expect growth hormone treatment to slow down aging and to increase longevity. Studies with animals, however, have not supported this idea in any convincing manner. Although the administration of growth hormone in old rats and mice has been reported to increase the weight of the muscles, liver, heart, kidneys, thymus and spleen; there is little evidence that it increases their longevity. Indeed, as mentioned earlier, hypophysectomy (removal of the pituitary gland) which abolishes the production of several hormones including growth hormone, has been shown to increase longevity. Despite this, growth hormone has been administered to old humans with a considerable degree of success in improving fitness and health. For example, Daniel Rudman and his colleagues working in the USA published a paper in the *New England Journal of Medicine* in 1990 that looked at the effects of growth hormone injections in 21 men between the ages of 61 and 81 years who had naturally low levels of this hormone. The results showed that the administration of growth hormone for 6 months was accompanied by an 8.8% increase in body mass; a 14.4% decrease in fat tissue, a 1.6% increase in lumbar bone density, and a 7.1% increase in skin thickness. According to Rudman these effects were equivalent to the changes normally seen during 10–20 years of aging. A number of other studies have looked at the effects of growth hormone in humans, including groups of athletes and young people, and broadly similar findings have been reported. Unfortunately, growth hormone treatment is costly – approximately $13,800 per year per person. There are also possible health risks including edema (swelling), hypertension and acromegaly (enlargement of the bones and excessive tissue growth). Nevertheless, in spite of its potential drawbacks, this work shows that with careful administration of growth hormone, aged body composition can be transformed into a seemingly more youthful state.

Whether growth hormone can result in life extension is far less certain. In humans a congenital lack of growth hormone does not necessarily mean a short life, since dwarfs have been shown to live well into their 70s, and possibly longer. Furthermore, hypertension and acromegaly are know to be serious health risks of increased levels of growth hormone. In one study that examined 194 acromegalic patients who had damage to their pituitary gland, it was found that their death rate was twice as high than normal people, and that they had a 4-fold greater risk of hypertension and heart disease. Thus, there are potential dangers with growth hormone therapy, and it is probable that it could speed up the progression of heart disease and cancer. Nevertheless, since growth hormone can apparently restore a more youthful appearance to older individuals, it could be that the benefits of this treatment far outweigh the risks in healthy individuals who have low levels of this hormone.

THE PINEAL GLAND AND MELATONIN

Not all endocrine structures come under the direct control of the hypothalamic-pituitary system, and one such example is the pineal gland; a structure which is currently attracting a great deal of interest and controversy for its role in aging. The pineal gland has been known since ancient times (it was allegedly discovered by Herophilus in 300 BC), although its function has only begun to be understood in recent years. In humans, the pineal gland is a single small white structure located deep in the brain, shaped like a pine cone after which it gets its name. Because of its distinctiveness and its single body, the philosopher Descartes (1596–1650) concluded that the pineal gland was the part of the brain where the soul resided. In fact, it wasn't until the end of the 19th Century, when it was discovered that pineal damage could produce precocious puberty and gigantism in children, that it began to be seen as an endocrine organ. Even then, many scientists were unconvinced because the pineal gland was shown to have a rich neural innervation which was at odds with the widespread belief that endocrine glands were only responsive to hormones in the bloodstream. It is now known that the pineal gland is an exception to the rule and that it acts to 'transduce' neural information into endocrine secretion, which in this instance is the hormone melatonin.

The pineal gland serves a number of important functions. In 1918, the Swedish anatomist Nils Holmgren was surprised to find that the frog pineal contained cells that bore a marked resemblance to the cells of the eye's retina. On the basis of this observation he suggested that the pineal gland might function as a photoreceptor or 'third eye' in cold blooded vertebrates. However, in most mammals, including humans, this idea made little sense because the pineal gland was located deep within the brain. But, Holmgren was not far from the truth since later research carried out in the early 1970s showed that the pineal gland received input from the visual system, via a small region of the hypothalamus called the suprachiasmatic nucleus (SCN), and that this visual input directly controlled melatonin release. In short, it was shown that darkness (or dim light) stimulated the release of melatonin from the pineal gland, and that bright light quickly inhibited its release.

In humans, melatonin follows a clear a circadian rhythm with very low levels during the day and a large surge occurring in the evening. However, in contrast to other animals, it appears that this rhythm is not directly caused by changes in light intensity; but is instead controlled by neural activity in the SCN, which is one of the few brain structures able to generate its own circadian rhythm. Therefore, in humans, it is the SCN which times the release of melatonin, at least under most circumstances. This is not to say, however, that humans have escaped completely from the effects of light on pineal secretion – they haven't, but it normally takes very strong light (some 4–5 times brighter than house light) to suppress melatonin release.

It has been difficult to attribute a precise function to melatonin. In animals it is known to control aspects of seasonal behaviour, including reproductive status, body weight and coat colour. And in children, pineal tumours have been shown to produce precocious sexual development and behaviour. Exactly what melatonin does in adults, however, is still far from clear. One effect of melatonin administration is to produce drowsiness and therefore one of its functions may be to initiate sleep. But, this may be part of a much larger involvement in the control of the body's circadian rhythms. Nearly every bodily function follows a 24 hour pattern of activity. To give one example, body temperature is at its lowest just before waking, rises to a peak around 9am and then falls rapidly throughout the day. Underlying and directing our bodily rhythms are circadian changes in the release of our hormones. For example, thyroid stimulating hormone is released in its greatest quantity during late evening and reaches its lowest level during the morning. In contrast, the peak output of cortisol occurs at around 9 am, and this level drops continually throughout the day until it reaches its lowest level around midnight. These rhythms persist even when the individual is cut off from all external cues and kept in constant light or darkness. In such situations circadian rhythms are believed to be driven by biological clocks in the brain such as the SCN (and there may be other clocks although these have yet to be identified).

It is thought that melatonin exerts an important circadian influence on the control of the hypothalamus, and to a lesser extent on the pituitary gland. Melatonin also directly affects many cells and tissues in the body and possibly makes them more receptive to circadian stimulation from other hormones. It appears, therefore, that melatonin plays an important role in synchronising the many hormonal rhythms of the body, particularly sleep; and it also provides an endocrine back-up to the SCN which has direct neural influence over the circadian release of hormones from the hypothalamic-pituitary axis. Despite this, much more remains to be learned about the function of melatonin in humans.

THE PINEAL GLAND AND AGING

The pineal gland reaches its biggest size in humans at around 4 years of age and then remains fairly constant in size into adulthood and old age. This is not to say, however, that the pineal is resistant to age-related changes. For example, with aging it builds up fibrous tissue, and there is also an increase of pigmentation in its cells. In addition, as the pineal ages it tends to calcify and becomes hardened. Although not proven, it would be surprising if the accumulation of these calcium deposits did not interfere with melatonin secretion in some way. In addition, it is also interesting to note that the pineal gland tends to be slightly bigger in women than in men.

The question of whether melatonin secretion changes with aging is made difficult by the fact that it is released mainly at night. Despite this, there is clear evidence that melatonin secretion does indeed decline with aging. For example, Franz and Maria Waldhause working at the University of Vienna have examined nocturnal melatonin levels in 280 individuals, aged from 3 days to 90 years, and found that melatonin levels were low in the first few months of life, increasing to reach peak values between 1 to 3 years of age, and then dropping rapidly and progressively until the end of adolescence (in fact, levels may decrease by some 75% over this period). The secretion of melatonin then declines very slowly throughout adulthood, perhaps falling by a further 10% over the next 50 years or so. Further research has also shown that in elderly people, the evening peak level of melatonin occurs later, and the surge tends to be reduced in amplitude resulting in a shorter duration of elevated nocturnal melatonin. In other words, aging is associated with a reduction of melatonin release and a flattening-out of its circadian rhythm.

Beginning in the late 1980s, Walter Pierpaoli and William Regelson working in Ancona, Italy, began to examine the effects of adding melatonin to the evening drinking water of groups of young and old mice. In their initial study they found that the administration of melatonin had a dramatic effect on the longevity of their animals. In brief, if melatonin was given to mice starting at about 18 months of age (this age was chosen by Pierpaoli because he estimated it to be the human equivalent of 65 years), then they lived on average 6 months longer than mice not given melatonin (931 days compared to 752 days). According to Pierpaoli this was a 20% increase in life span, which if applied to humans would amount to an extra 25 years of life. Even more striking was the general condition of the melatonin treated animals, perhaps best described in Pierpaoli and Regelson's own words:

The mice on melatonin had actually grown more fur and continued to boast thick, shiny coats. Their eyes were clear and cataract-free, their digestion had improved, and instead of growing thin and wasted in the manner on the non-melatonin-treated mice, they maintained their strength and muscle tone. The vigour and energy with which they moved around their cage resembled the behaviour of mice half their age.

These findings prompted Pierpaoli and Regelson to look at the effects of melatonin starting at different ages in different genetic strains of mice. Again, the results of these experiments confirmed the initial finding that melatonin given halfway through life significantly extended the life span of mice. However, if melatonin was given to younger mice, starting at about one year of age, no increase in life expectancy was found and such mice were also shown to exhibit a higher incidence of tumours as they got older. Thus, the beneficial effect of melatonin depended upon the age at which it was first administered. One of the most interesting findings from this work, however, concerned the effects of melatonin in New Zealand Black (NZB) mice; a strain that have a high degree of autoimmune disease in which the immune system attacks the tissues of its own body. These mice show early muscle wasting and have a greatly reduced average life expectancy of less than 18 months. However, in NZB mice, melatonin administration started at 4 months of age was shown to increase the life expectancy of these animals by up to 6 months. Thus, not only did melatonin increase longevity, but it also appeared to have a beneficial effect on the functioning of the immune system.

These findings provided evidence for the theory that the decline of melatonin secretion later in life was related to aging. More importantly, it also raised the possibility that the pineal gland might also act as a biological clock for aging in addition to its role as a timekeeper for controlling circadian rhythms. But how could this hypothesis be put to the test? One way, reasoned Pierpaoli, was to take pineal glands from young mice and to transplant them into old mice. Clearly, if this operation had a rejuvenating effect on the aged recipient then this would be very strong evidence for an involvement of the pineal gland in aging. Pierpaoli set about testing this hypothesis by removing the pineal glands from mice who were 3 to 4 months old, and then implanting them into mice of various ages between 16 and 22 months. Because of the difficulties in transplanting pineal glands, Pierpaoli implanted the new pineal into the thymus gland (located in the neck) and left the old pineal in place. The results nevertheless showed that the mice given pineal transplants lived on average 3 months longer than those that received no transplants.

Although this was a significant increase in longevity, this increase was not as marked as that obtained when melatonin was added to the mice's drinking water. One possible explanation was that Pierpaoli had not removed the old pineal gland. Thus, the old gland may still have been contributing to the animals aging. To examine this possibility in more detail, Pierpaoli along with the Russian researcher Vladimir Lesnikov working in St Petersburg, decided to undertake the much more difficult procedure of removing the pineal gland from mice (which is no bigger than a full stop at the end of this sentence) and inserting a new one into its place. During a series of operations they did this by transplanting young (4 month old) pineal glands into old (18 month old) mice, and in return, implanting the old pineal glands back into the young mice. According to Pierpaoli the results were remarkable. Within a few months both groups of animals looked to be the same age. The young mice visibly aged as a result of their operation, whereas the old mice had apparently become rejuvenated. Furthermore, the long term results showed that the young mice given old pineal glands died 3 months earlier than average (at about 27 months or 30% earlier than normal), whereas the old mice given new pineal glands lived 30%

longer than average, and died at about 33 months of age. In addition, it was reported that the animals transplanted with the 'new' pineal glands remained youthful looking and vigorous until the end of their life.

CRITICISMS OF THE MELATONIN HYPOTHESIS

These results, if they are reliable, would clearly be remarkable and have profound implications for the understanding of the aging process. Walter Pierpaoli and William Regelson have published an account of their work in the popular book *The Melatonin Miracle*, and this has become a best seller in the United States and attracted a great deal of media attention. Indeed, partly because of such publicity, melatonin has become big business, particularly in the USA, where it is widely available in health food shops and is said to be more popular than vitamin C. Thus, many people are already taking matters into their own hands and using melatonin to aid their health, and to help themselves live longer. This is despite the fact that no long term clinical trials of melatonin have ever been undertaken to guarantee its safety.

Despite all the media hype, much of the work undertaken by Walter Pierpaoli has been criticised by the medical community. For example, in his initial study looking at the effects of melatonin added to the drinking water of mice, Pierpaoli used strains of mice that had a genetic defect in the ability to produce melatonin. Whether Pierpaoli was aware of this is not clear, but nevertheless melatonin administration in his studies was probably just correcting a melatonin deficit that already existed in his animals. Indeed, in line with this interpretation it appears that melatonin does not increase longevity in normal animals, and more worrying, it may also be associated with an increased incidence of tumours, particularly of the reproductive organs. At the very least, the use of genetic strains of mice with defects in melatonin metabolism seriously distorts the claims made by Pierpaoli and Regelson regarding the anti-aging effects of melatonin.

A similar criticism can be made of Pierpaoli's work involving transplantation of pineal glands. Again, the genetic strains of mice used in this work were not the most suitable ones; and some researchers, such as Steven Reppert and David Weaver of Harvard Medical School, have gone as far as to say that the theory that young pineal glands transplanted into old mice keeps them young because of 'youthful' melatonin levels is absurd. In fact, considering the difficulty involved in transplanting pineal glands, it is somewhat surprising that Pierpaoli and Lesnikov did not undertake the simple procedure of actually measuring melatonin levels in their animals to see if the grafting had actually 'worked'. Indeed, the pineal gland, as we have already seen, is a somewhat unusual endocrine gland since it requires neural innervation (largely from the SCN) to control melatonin release. It is highly doubtful whether Pierpaoli and Lesnikov's grafts become innervated neurally, and they certainly did not attempt to see whether any such innervation occurred. In short, it seems unlikely that the transplanted glands could have produced melatonin in any reasonable amount.

Perhaps the most simple test of the pineal hypothesis is to examine the effects of pineal removal on aging. If melatonin is life enhancing and reduces aging then clearly pinealectomy should reduce the life span of laboratory animals. However, this does not appear to occur. Although pinealectomy may increases the circulating levels of fats and lipids in old animals, and possibly lead to greater tumour growth, there is no evidence that this operation significantly shortens life span. In fact, removal of the pineal gland doesn't even appear to disrupt the body's circadian rhythms, which casts some doubt on its proposed timekeeping role. Thus, much of the evidence suggests that the pineal's exalted role in aging may not be so special after all.

Despite all this evidence against the melatonin theory, Pierpaoli and Regelson begin the preface of their book by stating:

In this book, we will tell you how melatonin can help you 'grow' young, remain disease free, retain your sexual activity, get a better night's sleep, and enhance the quality of your life in a variety of ways.

Sadly, most of these claims, based on the evidence presented by Pierpaoli and Regelson, are not supported by convincing data, and therefore these dramatic claims have to be viewed with a certain deal of scepticism. This is not necessarily to say that they are false, but there is simply not enough evidence to make such bold claims with any degree of certainty. Clearly, much more research is needed but unfortunately such highly publicised claims tends to put off other scientists from looking more closely at the subject. This is reflected by the fact that many recent texts on aging, including books devoted to the hormonal basis of aging, and all three volumes of the *Handbook of the Biology of Aging*, make little or no reference to the pineal gland or melatonin. This is a great shame because claims for a pineal involvement in aging are probably just as strong as for most other hormones. However, as we shall see in the next chapter, melatonin is not the only hormone that has been the centre of extravagant claims. Much the same has been said about the sex hormones, particularly testosterone, except in this case the research went one step further with the development of human testicular grating as a means to slow down aging.

SUMMARY

- The endocrine system consists of a number of glands scattered throughout the body that secrete chemical substances called hormones (from the Greek *hormon* meaning to 'excite').
- The master gland of the endocrine system is the pituitary whose main function is to control the release of hormones from other glands of the body. In turn, the pituitary gland comes under the control of a brain structure called the hypothalamus.
- Removal of the pituitary gland in laboratory animals (which severely reduces the levels of many hormones in the body) has been shown to increase their life span providing they are kept under non-stressful conditions.
- Thyroid hormones are particularly important in regulating the body's metabolism although there is little evidence to show that levels of thyroid hormones decline significantly with age.
- Donald Denckla has argued that the pituitary gland releases a 'death hormone' which interferes or blocks the action of thyroid hormones on cells. However, this hormone has never been identified.
- Although it is far from certain, some evidence has shown that levels of glucocorticoids (which are released from the adrenal cortex often in response to stress) may increase in the aged. This could have serious repercussions since it is known that high levels of glucocorticoids are a potential cause of brain damage.
- Levels of the adrenal hormone dehydroepiandosterone (DHEA) declines significantly over the life span so that by the age of 70 it may only be 10% of its original youthful level. Low levels of DHEA have been found to be a significant risk factor for heart disease, and death *from any cause* in men over 50 years of age.
- Levels of growth hormone (which is released directly by the anterior pituitary gland) also decline with aging. In studies that have looked at growth hormone treatment in men, it has been claimed that it increases muscle mass and decreases fat tissue – equivalent to the changes normally seen in 10–20 years of aging.
- The pineal gland is an endocrine organ that is found in the brain and releases the hormone melatonin which is believed to be involved in

regulating biological rhythms. Melatonin is released mainly during the evening and the night, and its levels appear to decline slowly with aging.

- It has been claimed that melatonin administration (or pineal transplantation) in laboratory animals increases their life span and, in part, this has led to a growing media interest in the hormone and its widespread use. However, it may be the case that many of the claims made for this hormone are exaggerated.

SEX AND AGING

*The feebleness of old men is in part due to the
diminution of the function of the testicles*

–Charles Brown-Sequard

*The history of science, like the history of all human ideas, is a
history of irresponsible dreams, of obstinacy, or error. But, science
is one of the few human activities – perhaps the only one – in which
errors are systematically criticised and fairly often, in time corrected.*

–Karl Popper

The relationship between sexual vigour and youthfulness has fascinated mankind for thousands of years. It has long been recognised, not least by old men wishing to turn back the tides of time, that the period of youth with its benefits of fitness and attractiveness coincides with peak sexual performance. This relationship also makes a great deal of sense from an evolutionary perspective, particularly if one assumes that the purpose of life is to pass on one's genes through the process of reproduction. In most cases in the animal kingdom, this is only accomplished by the 'fittest' individuals and, as we saw in chapter 2, such competition helps guarantee the survival of the species. Taking the evolutionary theory one step further, it can also be argued that there is little need for an individual to survive past their reproductive (or rearing) period; and in the majority of species natural or accidental death does indeed occur before this point is reached. Although humans are fortunate in being able live past their capacity to reproduce (the only known member of the female gender to live past her period of reproductivity is apparently the human female) it is also clear that beginning around the peak of sexual maturity, our biological capacities decline at a constant rate (see chapter 1), and this includes sexual performance. It is hardly surprising, therefore, that sexual behaviour is inextricably linked with youth, and because of this association, numerous attempts have been made throughout the ages to postpone aging by methods designed to induce sexual rejuvenation. Our present century is no exception – although it must be said that the area has also attracted more than its fair share of quacks and charlatans whose presence still casts a controversial shadow over aging research. Consequently, the subject remains stigmatised despite the fact that science has made great strides in recent years in stripping away the myth and magic surrounding one of the most ancient (and still widely believed) theories of aging.

HISTORICAL ANTECEDENTS

From earliest times, the male testes have been linked with sexual vigour, longevity and bravery. It is known, for example, that the ancient Greeks and Romans used concoctions made from goat and wolf testes as aphrodisiacs and stimulants, and those who used such extracts included Caligula, Nero, and Messalina (the notorious wife of Emperor Claudius who used them to encourage her lovers). Similarly, testicular preparations have been found in many ancient medical texts, including those originating from the Middle East and China. The use of sex gland therapy also took place in the Middle Ages. For example, the famous Swiss physician Paracelsus (1493–1541) approved of sex gland therapy in the treatment of "imbecility", and the use of testicular extracts was fairly widespread in the treatment of impotence in Western medicine up to the 18th Century. Today, there are still many parts of the less developed world where such preparations, especially if taken from rare or powerful animals, are widely used to treat sexual problems, and undoubtedly furtive use of such extracts also continues in more advanced societies as well.

For thousands of years the effects of castration on both humans and animals has been documented, linking the function of the testes with masculinity and sexual vigour. But the first person to suggest that the testes may exert these effects by secreting a special chemical into the blood was probably the French physician Theophile de Bordeu, who in 1749, proposed that the sex organs imparted masculine or feminine characteristics to their hosts. It took another hundred years, however, before experimental proof of this idea was provided by Swedish scientist Arnold Berthold who is credited with performing the first formal experiment in endocrinology. Berthold removed the testes of two roosters and found that they no longer crowed or engaged in sexual or aggressive behaviour. When Berthold transplanted the testes back by inserting them into the abdomen, he found that sexual and aggressive behaviours were quickly restored. Later examination of the testes showed that although they had not reestablished nerve connections, they had nevertheless become attached to the intestine where they acquired a good blood supply and were in good condition. The obvious conclusion was that the testes must be secreting a chemical into the circulation that was determining the temperament and sex characteristics of the birds. We now know this chemical to be the male sex hormone testosterone.

Despite Berthold's pioneering work his findings lay forgotten for many years, and the idea that bodily secretions could somehow control temperament and other biological processes remained unproven. In fact, it wasn't until 1902 that the first "internal secretion" (called appropriately enough 'secretin') was discovered, and soon after the word 'hormone' found its way into the English language. The question of whether the testes were able to impart significant effects on the body around this time, however, was still highly contentious. Much of this controversy was due to the highly publicised work of one of the modern fathers of endocrinology – the French physiologist Charles Brown-Sequard. In many respects the field of aging research has never quite recovered from his proclamations regarding the effects of testicular extracts made to the Biologie de Societe in Paris, on 1st of June, 1889.

CHARLES BROWN-SEQUARD

Charles Brown-Sequard (1817–1894) was one of the most famous biologists of his time. The imposing 6'4" bearded professor was the author of over 500 research articles and had held many important medical research posts in France, England and America. In 1856, he had become the first scientist to show that removal of the adrenal glands resulted in fatal consequences, and he also undertook important work on spinal cord regeneration and epilepsy. In 1878 (at the age of 61 years) he reached the pinnacle of his career by being awarded the highly prestigious chair of experimental medicine at the College de France. For most of his life Brown-Sequard had been a very prolific researcher, but as he approached his seventieth birthday, he became increasingly aware of a decline in his ability to concentrate and work. He started to be easily fatigued and noted a decline in his muscular strength. Despite being recently married with a young wife, he experienced waning sexual vigour, and his nights were disrupted with sleeplessness that left him feeling tired and unrefreshed. Obviously worried about his declining vigour, Brown-Sequard began to think about ways of rejuvenating himself.

Brown-Sequard believed in the theory that the testes, or more specifically the seminal fluid that contained the sperm, had invigorating properties, and at the age of 72 years he began to test the effects of these extracts upon himself. He was clearly aware of the traditional beliefs concerning this type of therapy, but his views were also in accordance with the popular Victorian notion at the time that loss of semen through sexual intercourse (or worse still, through the practice of masturbation) led to a loss of strength and vitality. It sounds quite absurd now, but in the last century sexual indulgence and masturbation were widely believed to lead to mental illness, imbecility, disease and even death. Because of this danger, men were routinely advised by their doctors to conserve their sperm and to engage in sexual behaviour infrequently. This belief was perhaps best summed up by the American Orson Fowler in the 1870's who wrote of over indulgence:

It not only poisons your body, destroys your rosy cheeks, breaks down your nerves, impairs your digestion ... it corrupts your morals, creates thoughts and feelings the vilest and worst possible, and endangers your very soul's salvation.

Against this backdrop of feverish opinion, it is perhaps not surprising that Brown-Sequard believed that the testes produced an active ingredient that was essential to the health of the body, and whose decline led to the mental and physical deficiencies found in old men like himself.

In his first attempt to rejuvenate himself, Brown-Sequard ground-up the testicles of a young dog with a pestle and mortar, and then passed the mixture (which included semen and blood) through filter paper. He then mixed the tissue in distilled water and injected it directly into his leg. The beneficial results were almost immediate and after a few injections, Brown-Sequard found himself feeling significantly stronger and rejuvenated. Shortly afterwards he repeated the procedure twice more, this time with extracts taken from guinea pigs with similar results. In his own judgement the improvements in his well-being were remarkable and on 1st June 1889, Brown-Sequard made the bold step of presenting his findings directly to an audience of distinguished scientists at the Societe de Biologie in Paris.

Brown-Sequard began his presentation by talking about the general importance of testicular secretions for well being and the probability that their decline was linked to aging. He went on to describe his own aging decline and how this had led him to experiment upon himself with testicular tissue. Holding a small phial of testicular extract in his hand, he then described to his audience in explicit and vivid terms the effects of his experiences.

After just eight injections I felt much better, and more like my old self or a man half my years. I have regained the strength I used to have, and can carry out experiments continuously for several hours without needing to sit or rest. For some years I had been unable to do any serious mental work in the evenings, but on 23 May, after three and a half hours of continuous laboratory work, I felt so alert and energetic that after dinner I was able to write for nearly two hours on a difficult subject. You can imagine how the return of my old energy has lifted my spirits.

The benefits are not only mental. I had lost the ability to hold much water in my bladder, and the flow of urine was little more than a trickle. When I measured the jet in the pissoir again after the injections, I found that its length had increased by one third!

Constipation is even more of a misery. The muscles of the bowel, like those of the bladder, are controlled by nerves in the spinal cord, and as nervous activity declines so does muscular strength. After a few days of treatment, however, the power and regularity of my bowel has improved more than any other function and I no longer need laxatives. The tonic evidently improved all aspects of my spinal cord functions.

For scientists like ourselves, an improved feeling of well-being is not enough to prove that a real organic change has taken place – we need objective facts and figures. I have therefore used a dynamometer to provide an accurate measure of my forearm strength. Before my experimental trial, the average weight I managed to move was about 34.5 kilograms, but afterwards I lifted 41 kilograms. This is the weight I used to be able to lift when I lived in London twenty six years ago. The result implies the treatment was improving my muscles as well as my nerves.

I don't ask you to accept my claims uncritically. In your place, I would be among the first to question whether the good results were due to auto-suggestion rather than an organic change. I cannot yet be absolutely certain, but the good outcome was so unexpected that I doubt this interpretation. Confirmation is urgently needed by independent researchers. I challenge members of the society to test for themselves whether testicular fluid possesses rejuvenating powers.

Brown-Sequard added extra spice to the talk by euphemistically admitting that he had been able to pay his wife "a visit" for the first time in a long time. The talk lasted barely 15 minutes but its impact was enormous. Because of Brown-Sequard's tremendous prestige and the highly personal nature of his revelations, the talk was reported and sensationalised in the national press, and it provoked much debate in scientific circles. In a remarkably short space of time, Brown-Sequard had become the world's foremost advocate for the existence of "internal secretions" or what we now know as hormones. In this respect Brown-Sequard was soon to be vindicated but, in regards to his views concerning testicular extracts and rejuvenation, history was to treat him much less kindly.

The publicity that Brown-Sequard's talk generated ensured that there was an instant demand for his extracts – from fellow scientists and doctors ready to test the new miracle cure, and from the wider public eager to rejuvenate themselves. The demand was so great that French national newspaper *Le Matin*, funded, largely through the donations of its readers, an Institute of Rejuvenation where aging Frenchmen could obtain the new fountain of youth (in this case filtered bull testes). Brown-Sequard also provided free of charge his extracts to medical practitioners, and it has been estimated that within a year of his talk, some 1,200 physicians world-wide were testing his elixir in clinical trials.

Unfortunately, subsequent research was unable to duplicate Brown-Sequard's findings, and soon there was considerable scepticism and vociferous criticism. By the late 1890s the use of testicular extracts were on the decline, and the ground swell of public opinion appeared to support the medical view that the therapy was ineffective. One German medical publication wrote that "his fantastic experiments...must be regarded as senile aberrations", and another wrote that Brown-Sequard was "proof that professors should retire at the age of threescore years and ten". Perhaps the strongest criticism came from Britain, with the Editor of the *British Medical Journal* stating "We find medical men writing of these ideas and of the cures achieved... and often upon no better evidence than quacks produce for their cures". Brown-Sequard was an easy target for ridicule and his professional reputation was soon irrevocably damaged by the saga. The last few years of his life were particularly difficult for Brown-Sequard. His own health began to deteriorate, his wife left him for a younger man, and despite keeping up his injections, he died from a stroke in 1894 at the age of 77. The end of Brown-Sequard was also the end of his testicular extracts and the medical establishment quickly distanced itself from the treatment.

Why had Brown-Sequard been proven so wrong in his strong belief of the effectiveness of testicular extracts? This question is still not easy to fathom. We now know that the testes do indeed produce a hormone (testosterone) and therefore Brown-Sequard was correct in his speculation that the testes released a special substance into the blood. It was probably the case that his extracts contained a small amount of this substance, although most biologists believe that it would have been hardly enough to produce a biological response. Moreover, if any effect had been obtained, it would have been marginal and short-lasting. There are several possible reasons for Brown-Sequard's mistaken conclusions, but perhaps the most plausible relates to what is known as the placebo effect (or what Brown-Sequard himself referred to as "auto-suggestion"). In short, it has been suggested that Brown-Sequard was expecting an effect, and perhaps his optimism over the likelihood that his injections were going to have the desired result, produced an effect in his mind that was not matched in his body. This type of bias is extremely common in any type of research involving psychological interpretation, and for this reason most modern drug research is considered highly suspect, if not worthless, without control groups that include participants who do not know what treatment they are being given.

Despite its sensationalism and personal tragedy, many historians nevertheless see Brown-Sequard's talk as the modern starting point of the science of endocrinology. Because of his impetus, extracts of virtually all animal tissues were tested rapidly in the 1890s and soon genuine and revolutionary discoveries were being made. In 1891, the British doctor George Murray was the first person to successfully treat a woman with thyroid disease using sheep thyroid extract, and in 1893 George Oliver discovered that adrenaline (obtained from the adrenal glands) had a stimulatory effect on blood pressure. Brown-Sequard had been one of the first scientists with the foresight to predict the existence of these "internal secretions" and in this respect he has been proven right.

The identification of the active testicular ingredient, however, remained elusive and was only finally discovered in the 1930s when new chemical techniques were brought to bear on the problem. In 1931, the first masculinising hormone was extracted and crystallised from 25,000 litres of urine collected from Berlin policeman. It was called androsterone and was found to be a steroid-type substance derived from cholesterol. It soon became clear, however, that androsterone was not the only substance being secreted by the testes, and in 1935 a much more biologically potent substance was discovered by a Dutch team working in Amsterdam using bull testes. This was the hormone testosterone, and at long last the active and allegedly rejuvenating substance that Brown-Sequard had sought, had been finally discovered.

THE YOUTH DOCTORS

Although the use of testicular extracts was ridiculed by many leading medical authorities, Brown-Sequard was not the only doctor who believed in the link between sexual decline and aging, or who provided treatment that promised rejuvenation based on such ideas. In fact, a number of medical practitioners attempted to follow in his footsteps. Although some have been fakes and fraudsters, others have been prestigious and highly respectable scientists who sincerely believed in the treatments they discovered. The issue has never quite gone away, and the story of Brown-Sequard would not be complete without looking at some of the main figures he influenced (sometimes called the 'erector set') who have also tried to turn back the tide of aging by offering sexual rejuvenation.

EUGEN STEINACH

In 1890, Eugen Steinach (1861–1944), professor of physiology at the University of Vienna, influenced by the work of Brown-Sequard, began to examine the effects of grafting young testicles and ovaries into old animals. His operations were successful and he reported that his old male rats were reinvigorated and lived up to 25% longer with the implantation of new testicles; whilst old female guinea pigs given new ovaries were apparently able to conceive and become pregnant once again. These were dramatic findings and the next logical step was to implant humans with sex glands. However, Steinach was reluctant to pursue this more controversial path, and he sought simpler solutions to the problems of sexual rejuvenation. In fact, he was much more interested in producing rejuvenation by boosting existing hormone production from old testes, but the question was how to do it?

The answer came from a somewhat unusual source. Biologists had long been aware that certain plants could be kept in a perpetual juvenile state by removing their reproductive parts as they appeared; and this finding had been used to support the argument that energy diverted from reproduction could be used instead for the health and maintenance of the body resulting in a longer life (see also page 49). Using a similar argument, Steinach reasoned that if the sperm-producing function of the testes was inhibited, then the glands would increase testosterone production by way of compensation, and this would have rejuvenating effects. In 1920, Steinach tested this theory by severing the vas deferens (the tube carrying sperm from the testes) in his laboratory rats. The results of this operation showed that these animals not only lived longer, but they also appeared to be in better health. In other words, the secret of youth appeared to lie with an operation now widely used to bring about male sterilisation – that is, a vasectomy!

It was not long before Steinach applied his research to human patients and in his view with equally successful results. Amongst the improvements that the operation apparently provided were increases in physical strength, energy and mental performance; a decrease in blood pressure; and perhaps most valued of all, increased libido. Such spectacular claims were not only made by Steinach, but by many other respectable doctors who used the technique. Indeed, because of the initial widespread testimony to its effectiveness, the "Steinach rejuvenation operation" became a popular form of surgery with a host of famous people undergoing the procedure including the Irish poet W.B. Yeats. The general consensus seemed to be that many patients were delighted with the operation and Steinach's reputation grew immensely as the positive reports flooded in.

However, by the 1930s doubts were beginning to be raised about the effectiveness of the operation, and the weight of mounting medical opinion started to swing against Steinach. It gradually became clear that many people did not receive long term benefits from the operation, and that a placebo effect probably accounted for the people who did. Enthusiasm waned for the technique and it quickly became obsolete as more and more critics spoke out against it. Today, the procedure is regularly performed as a means of sterilisation, and it is interesting to note that nobody is now heard to say how reinvigorated they feel afterwards. This is perhaps the best reason of all to reject Steinach's claims and, as with Brown-Sequard, it is hard to escape the conclusion that the success of the operation was founded on the power of suggestion.

SERGI VORONOFF

Although Steinach had transplanted testes (and ovaries) into animals, he had not done so with humans. However, there were others willing to try this operation, and the person most famous for taking up the challenge was a Frenchman of Russian aristocratic stock called Sergi Voronoff (1866–1951). At the turn of the century, Voronoff was physician to the Egyptian ruler Prince Abbas, and as part of his duties he was responsible for the medical treatment of members of the Prince's royal court, including the eunuchs that guarded the King's harem. Voronoff soon realised that he was spending a disproportionate amount of his time treating these castrates with medical complaints that appeared to be linked to their tendency to age much more quickly than usual. Because of this experience, Voronoff, in line with Brown-Sequard and Steinach, came to the conclusion that the testes was secreting a special substance with anti-aging properties. On his return to France in 1910, Voronoff set about testing his theory by transplanting slices of young testis gland into castrated rams, and into animals that had reached old age. Voronoff reported spectacular success with his methods, and claimed that he was able to restore normal sexual activity, weight and horn growth to castrated animals, and to revitalise the older rams. These findings were also repeated with other kinds of animal including bulls and sheep. Within ten years of starting his research, Voronoff was confident enough to begin undertaking the operation on his first human patients.

In fact, Voronoff was beaten to the accolade of being the first surgeon to perform testicular transplantation in humans. On 18th October 1919, in San Quentin prison, California, a convicted murderer named Thomas Bellon was executed, and his testicles immediately removed and transplanted into a prematurely aged inmate of the prison by Dr Leo Stanley. This doctor was to perform over 30 similar operations and to publish his findings in the respected journal *Endocrinology* where the work was described as a great success. However, Voronoff was not far behind, and on 12th June 1920, he carried out the first ape-to-man testicular extract operations in two patients who had both lost their testicles as a result of tuberculosis. Although the results from these first two operations were to be disappointing, the next two men to receive testicular grafts from Voronoff showed significant rejuvenation, one of whom claimed that the operation made him feel 20 years younger. Blocked in his attempt to present his results at reputable medical conference, Voronoff retaliated by giving a press conference where his findings provided a sensation. Suddenly, all the world knew about the invigorating benefits of his testicular transplants, and soon there was great demand for this new type of surgery. Over the next seven years, Voronoff was to privately implant chimpanzee testicles into at least 1,000 old men at the cost of between £500 and £1,000 – an astonishing fee for the 1920s. The financial rewards were huge and it has been estimated that Voronoff may have earned over $10 million from his lifetime's work involving this type of surgery.

Voronoff published full details of his surgical procedures and they make interesting, if somewhat gruesome, reading. His operation first involved removing the testicles from a young adult monkey and cutting each piece of tissue into six thin slices. The patient's scrotum was then opened under local anaesthetic, and the exposed testicle scraped to ensure good capillary contact, on to which each testicular slice was stitched into place using silk sutures. Voronoff reported that graft normally 'took' in a few days, when it presumably started to release its own secretions into the bloodstream. Following the operation, Voronoff's patients usually rested for a few days, although this was not insisted upon, and typically no ill effects from the procedure were reported.

As with Steinach, early reports on the effectiveness of the Voronoff technique were favourable and it received good publicity. However, during the 1930s, evidence began to mount showing that the operation was not as effective as was first believed. Many of the first negative findings came from veterinary studies in which Voronoff's operation was used in an attempt to produce livestock improvement, or to rejuvenate old animals for breeding purposes. These studies often used objective criteria to measure the results, and control (non-operated) groups of animals to enable meaningful comparisons. Such work convincingly showed that the operation produced no improvement in animals and, not surprisingly, this made other investigators take note and question more seriously the application of the technique to humans. The final irrefutable proof of the futility of the operation, however, didn't emerge until after the Second World War, when Peter Medawar began examining the possibility of using skin grafts to treat burn injuries. Medawar found that skin from one animal was nearly always rejected after grafting to another, and the more genetically dissimilar the two animals the quicker the rejection. Moreover, with the exception of identical twins and highly inbred strains of animals, the grafts was always rejected for the simple reason that the immune system recognised the new tissue as foreign, and consequently mounted an immune response that quickly destroyed it. In short, there was no way a transplant taken from a monkey and placed in a human could have taken hold and 'worked'. In fact, it is only over the last 25 years, or so, that immune suppression drugs have permitted the first successful transplants between unrelated humans. At best it was perhaps possible that Voronoff's transplants could have established a temporary connection with the hosts circulatory system and functioned for a short time, but even this was highly unlikely. Again, the power of suggestion, it would seem, provides the only possible explanation.

PAUL NIEHANS

The legacy of testicular transplants survives to this day in the form of "cell therapy" in which cells are taken directly from the tissues of foetal or new-born sheep, pigs, or rabbits and injected into human recipients. The person who first developed this technique beginning in the 1930s, and who arguably remains the most famous of all professional rejuvenators, is the Swiss Nobel Prize laureate Paul Niehans (1882–1971). The basic idea of cellular therapy is easy to understand. If a particular organ, or system of organs, appears to be malfunctioning, the cellular therapist treats it by injecting 'youthful' extracts of that particular organ taken from another animal. Alternatively, if the problem is caused by over-function of an organ, then treatment is made by injecting the extract from an antagonistic organ.

The first step in cellular therapy is to establish what organs are malfunctioning. To do this Niehans asked his patients to provide a urine sample which then underwent a rather obscure medical procedure called the Abderhalden test, which until recently could only be undertaken by its originator Emil Abderhalden (and then later by his widow). Following the results of this test (which were apparently able to pinpoint the hormonal failings of the patient) the person was then able to undergo treatment, involving a series of injections spaced over several days. Niehans claimed that to be maximally effective the injected cells should be prepared and injected within seconds, or at the very most minutes, after they have been removed from the donor animal. Tissues used by Niehans, depending upon the exact problem of the patient, included the placenta of sheep; the hypothalamus, pituitary gland and parathyroid from calves; the testes from young bulls; and ovaries and adrenals from young pigs. A suspension of fresh cells was then injected deep into the gluteus maximus muscle (otherwise known as the buttocks) with a separate injection needed for each type of tissue used.

Surprisingly, Niehans did not claim to understand exactly how his treatment worked. One possibility is that the cells injected into the muscle remain alive and migrate towards the appropriate and deficient endocrine organs resulting in their stimulation and rejuvenation. Alternatively, it may be that the cells stay alive at the point of injection, and this in turn activates the deficiencies of the degenerated organs from a distance. Whatever the mechanism, Niehans claimed to have obtained striking improvements in a wide range of disorders including hypertension, depression, nervous disease and diminished libido and impotence. The testimony to the technique has been so great that Paul Niehans has treated some of the world's richest and most famous people including Somerset Maughan, Noel Coward, Winston Churchill, the Duke and Duchess of Windsor and, perhaps most famous of all, Pope Pius XII who was seriously ill and received his first cellular therapy in 1935. Eight weeks after the treatment, the Pope was able to take up full duties again, and he lived for another 23 years, during which time he repeated the treatment on two more occasions.

It is difficult, if not impossible, to judge the effectiveness of cellular treatment objectively because it has not been seriously tested by the medical establishment. Most doctors assume that foreign material (i.e., the cellular suspension) injected into the body will result in an immune reaction that will quickly destroy the cells. However, there is a question mark hanging over this particular assumption since claims have been made that embryonic tissue has significantly fewer antigens than normal tissue, and is thus less able to invoke an immunological response. Indeed, although a dangerous immune shock reaction can sometimes take place following treatment, it appears that this is a rare event which supports these claims. One of the few attempts to measure the effectiveness of cell therapy was undertaken by Niehans himself in 1963. He sent out a questionnaire to all of his patients, and found that some 89% of his respondents were convinced of the effectiveness of the treatment. Despite the risks and lack of evidence, cell therapy remains a lucrative business in many countries including Switzerland, Germany, Holland and the Bahamas, although it is illegal in the United States and does not appear to be practised Great Britain. Hopefully, one day somebody will be able to judge objectively whether there is something to the technique, or whether it is just another example of man's seemingly natural tendency to fool himself.

TESTOSTERONE

The hormone allegedly at the heart of all these rejuvenation remedies is, of course, testosterone, which in males is secreted mainly by the testes (although small amounts are also produced by the adrenal cortex). The testes (or gonads) secrete several male hormones which are collectively known as androgens, but it is testosterone that is produced in the highest amounts and for this reason, it is considered to be the most significant masculinising hormone in the male. Perhaps because of its controversial history, it is also a hormone that many people associate with sexual decline and the aging process.

Any man who has ever been kicked in the groin must wonder why his testicles are situated in such a vulnerable location. The reason is simply that the testes are the site of sperm production, and for this process to occur the temperature of the testes needs to be about 2°C less than the rest of the body. The testes are essentially sacs of tightly coiled tubes (called seminiferous tubules) that produce sperm. In fact, each testis contains over 800 seminiferous tubules which produce thousands of new sperm every second in healthy young men. Remarkably, if all the seminiferous tubules from both testes were laid out in a straight line they would stretch some 225 metres! It is perhaps surprising that there is any space left in the testes with all these tubules, but there is, and situated amongst the seminiferous tubules are clumps of Leydig cells that produce testosterone along with the other androgens.

As with most other endocrine systems in the body, the release of testosterone from the Leydig cells is ultimately under the control of the hypothalamus and the pituitary gland. Testosterone production begins when the hypothalamus releases gonadotrophin releasing hormone (GnRH), which in turn stimulates the anterior pituitary gland to release two chemical messengers into the bloodstream, called luteinizing hormone (LH) and follicle stimulating hormone (FSH). These hormones are better known for their control of the menstrual cycle in women; but in men it is LH that stimulates the production of testosterone by the Leydig cells, while FSH stimulates sperm production from the seminiferous tubules (which also requires the presence of testosterone). The net effect of this stimulation is to increase the levels of testosterone in the blood. Increased levels of testosterone will in turn, feed back to the pituitary and hypothalamus to suppress further release of LH; whereas in the case of sperm production it appears that the seminiferous tubules produce a hormone called inhibin which inhibits the release of FSH.

The Effects of Testosterone

In general, the functional role of testosterone is to produce the distinguishing sex characteristics of the male body. Testosterone begins to exert its effects as early as 6 weeks after conception when it acts to determine the sex of the foetus. If this did not occur then we would all turn out to be females since it is testosterone that stimulates the prenatal growth of the male sex organs from the female blueprint, thereby determining our biological gender. In males, testosterone levels remain fairly high until a few months after birth, when secretion effectively shuts down until puberty. With the onset of puberty, however, testosterone production increases rapidly again, and sets into motion the secondary sex characteristics of the male. At this point testosterone produces a wide range of effects. For example, it stimulates the growth of body hair and enlarges the larynx resulting in a deepening of the voice. The skin gets thicker and becomes oilier (which makes young men susceptible to acne), bones grow longer, and muscles increase in size resulting in increased body mass. Testosterone also causes the penis to enlarge in size and to stimulate the testes into producing sperm. In short, testosterone turns a boy into an adult male.

Testosterone also continues to exert an important influence throughout life. One of its main actions is to stimulate the sex drive in both males and females, regardless of whether they are heterosexual or homosexual. Essentially, too little testosterone is associated with declining sexual interest whilst too much is associated with increased desire. It is also generally accepted that testosterone is linked with competitiveness and aggressive behaviour, and there is also strong evidence that it is even involved in certain types of thinking, such as finding one's way around an environment, or tasks requiring visual-spatial skills. Testosterone also boosts the metabolic rate and assists in protein formation. The mechanisms by which testosterone produces these effects in the body are complex, and may require it to be transformed into other chemical agents; but in the case of protein formation it is known that testosterone diffuses into the cells where it directly activates certain genes. Thus, testosterone helps to increase and maintain lean body mass and muscle.

AGE CHANGES IN LEVELS OF TESTOSTERONE

It is generally accepted that levels of testosterone tend to fall with age following puberty, although this decline is not as marked or predictable as most people think. For example, combining the results of 14 studies published in the 1980s, looking at levels of testosterone in people of various ages, it was found that old men had average blood levels of testosterone that were only 14% lower than their younger counterparts. In most cases the decline of testosterone appeared to occur around the age of 50, and then fall with the advancing years. However, there is great variance in the data, so that although some old men in their 70s to 90s have testosterone levels well below normal, others have normal or even high levels of testosterone. In fact, some studies have even found no decline in testosterone with aging. For example, the Baltimore Longitudinal Study (an ongoing research program undertaken by the National Institute of Aging, where subjects undergo extensive testing every couple of years, on a wide range of tasks designed to measure the results of aging) have regularly measured testosterone levels in 69 of their subjects over a long period. Surprisingly, this research has shown that there is no consistent decrease in testosterone, even at the highest ages. Although these subjects may not be representative of the whole population, the results show that old men in good health may have testosterone levels similar to that found in younger men. In other words, the decline in testosterone is not an inevitable feature of aging.

Despite this, there are other declines in the reproductive chemistry of the aging male. One distinct change is in the number of Leydig cells that produce testosterone. For example, a young male who is 20 years of age might be expected to have more than 700 million Leydig cells in his testes, but these then decline at a constant rate of approximately 80 million cells with each decade of life. In other words, by the time the male reaches 70 years of age, he has lost over half of his sex hormone producing Leydig cells. The consequences of this loss can be shown by injecting the older male with a protein called hCG which mimics the effects of LH and triggers testosterone production by the Leydig cells. In general, older men (aged 70–89) produce about 50% less testosterone in response to hCG than younger men (aged 25–49) regardless of their circulating serum levels of testosterone. In other words, the response of older men to hCG administration suggests that the surviving Leydig cells are working close to their maximum capacity to compensate for their reduced numbers.

An alternative way of measuring testicular function is to measure sperm production. In general, the findings here parallel those obtained with testosterone. That is, a declining sperm count tends to occur with aging, although this is highly variable and some old men appear to be particularly well endowed with sperm. A better measurement, however, may be the decline in seminiferous tubules. Around 90% of seminiferous tubules are manufacturing sperm in young adult males, whereas this figure can drop to 10% in men who are 90 years of age. Despite this, sperm is produced throughout life – presumably because the remaining tubules are working near to their full capacity. It is also interesting to note that in terms of world records, regular coitus with mutual orgasm has been recorded between a man of 103 and his wife aged 90; whilst the oldest age at which a man has fathered a child is recorded at 94 years.

DOES TESTOSTERONE INFLUENCE AGING?

As we have seen in this chapter, there have been a number of investigators who have staked their reputation on the use of testicular extracts, or their transplantation, as a means of slowing down or reversing aging. Since testosterone is by far the most important hormone produced by the testes then clearly, if these investigators are correct, we would expect testosterone to have rejuvenatory properties and its decline linked with aging. On first sight this does not appear to be a promising idea since, as we have already seen, testosterone levels do not necessarily decline with aging. But there may be an even better reason to be suspicious of the rejuvenatory effects of testosterone; that is, the simple fact that throughout the world, women nearly always live longer than men (see also pages 61–64). For example, in the UK the average life expectancy for a male is 74 years compared to nearly 80 for the female, and similar figures occur throughout the Western world. In addition, this relationship appears to hold true throughout the animal kingdom with relatively few exceptions. This is clearly not a good advertisement for the benefits of testosterone as a fountain of youth.

There are many reasons why females live longer than males, but it is probably fair to say that testosterone is one of the factors in the equation. One of the great benefits of being a female is the reduced likelihood of developing the most common cause of death found in industrial societies, namely coronary heart disease. If we look at the ages when heart disease starts to manifest itself, we find that men become particularly prone to heart problems after the age of 40, whereas women do not start show increased vulnerability until they reach their late 40s. (some 5–10 years later). Although the incidence of heart disease in females slowly catches up with male levels in old age (70–80 years), there is nevertheless a considerable gender gap in the numbers who finally succumb to this disease.

One of the reasons for this difference appears to lie with the sex hormones and their effects on cholesterol. Up until puberty, both males and females have equivalent blood levels of cholesterol, but following puberty the levels tend to increase in the male, largely as a result of testosterone. Even more important, however, is the fact that the type of cholesterol, or more accurately its 'carrier', also shows a difference between the sexes. The transport of fats (including cholesterol) around the body poses a problem because the blood is a watery medium, while fats are 'greasy', and the two do not mix together. Because of this, fats have to be transported in large protein molecules (that can be likened to supertankers carrying oil) called lipoproteins, of which there are two types: high-density lipoproteins (HDLs) and low-density lipoproteins (LDLs). In terms of heart disease lipoproteins can be regarded as good and bad. The HDLs are good because they travel through the blood soaking up excess cholesterol like a sponge. But, in contrast, the LDLs are the main carriers of cholesterol and, not surprisingly, they increase in people with diets high in cholesterol. They are bad because they have been shown to form hard plaques in the heart's coronary arteries leading to narrowing or blockage.

Returning to the sex differences that are known to exist in heart disease; it has been shown that testosterone tends to increase the levels of LDLs in the blood, whereas the female sex hormones (see below) increase the levels of beneficial HDLs, as well as also importantly decreasing the LDLs. This partly explains why heart disease is less common in females, although this advantage changes after the menopause when the female body shuts down its own supply of sex hormones. At this point the LDLs increase at the expense of the HDLs, and an increased risk of heart disease is the result. Why testosterone increases the levels of troublesome LDLs is not clear, although it is probably related to its tendency to increase metabolism. Whatever the explanation, it shows that there are health risks associated with testosterone, especially in middle age and beyond.

THE EFFECTS OF CASTRATION

An alternative way of examining whether testosterone influences the aging process is to examine the effects of castration on health and longevity. As mentioned earlier, Voronoff came to the conclusion, whilst serving as a young doctor in Egypt, that eunuchs had more health problems and appeared to age much faster than normal men. These observations were, of course, the springboard for Voronoff's subsequent work on testicular grafting. Despite this, there is little scientific evidence to show that castrates age faster than normal, although there are few studies to go by. For obvious reasons, castration is nowadays only performed under extreme circumstances, and the only present legitimate use of this operation is in cases of testicular or prostrate cancer.

Unfortunately, no studies have looked at the long term consequences of this operation on longevity. However, castration was performed in the earlier part of this century for somewhat different reasons. For example, in some European countries, castration has been used in the treatment of sex offenders, and in some states of the US castration has occasionally been undertaken in cases of extreme mental retardation. Although these treatments are, quite rightly, regarded as unacceptable today, they have nevertheless provided the only large groups of subjects by which to assess the effects of castration.

Despite its obvious interest to aging researchers, the only reported study that has looked at the effects of castration on longevity has been undertaken with mentally retarded patients. This study published by James Hamilton in 1969, compared 297 institutionalised mentally retarded men who had undergone castration for the treatment of behavioural problems during the years 1871 to 1932, with a group of 735 similarly retarded men who had not received the operation. Although the results may be confounded by the fact that mentally retarded populations tend to have shorter life spans than normal, the results showed that the castrated group lived on average for 69.3 years compared to 55.7 for intact males – a difference of nearly 14 years. Furthermore, similar findings have been reported in animal studies. For example, James Hamilton demonstrated that a group of rats castrated at birth lived on average for 526 days compared to 454 days for intact animals (an increase of about 14%). Similarly, Hamilton also reported that castrated male cats appear to have a longer life span than intact cats. And, as we have seen in previous chapters, pacific salmon and marsupial mice also benefit from the effects of castration. In general, it appears that castration may well have beneficial effects on male longevity.

Further support for this theory, although admittedly of a much less scientific nature, comes from examining the lives of some of the world's most celebrated opera singers who were castrated during the 1700s and 1800s. The life histories of many of the most famous castrati, complete with reliable dates of birth and death, are recorded in Angus Heroit's book *The Castrati in Opera* (1956), and an examination of 27 of these famous singers show that the average age of death was 65.73 years, and the longest lived individual was 84 years. Although these individuals obviously led privileged lives, their life spans are probably higher than expected for the times they lived in. At the very least it would seem that castration is not disadvantageous for longevity.

THE EFFECTS OF TESTOSTERONE ADMINISTRATION

Surprisingly, there has been little research that has examined the effects of testosterone administration upon aging. The one exception has been a study undertaken by Joyce Tenover at the University of Seattle, who looked at the effects of weekly testosterone injections in 13 healthy men aged 57 to 76 who had low (but within 'normal' limits) levels of testosterone. The study lasted for 6 months with subjects being given testosterone for a 3 month period, and a placebo injection for 3 months (subjects did not know what treatment they were given until after the experiment). The results showed that the hormone injection increased circulating levels of testosterone by nearly 80%, and during this time lean body (muscle) mass increased by an average of 1.7 kg, although there was no decrease in fat, or any change in the waist/hip ratio. Although the subjects were not told what injections they were being given, 12 out the 13 men were able to predict correctly the 3 month period in which they had received testosterone. Reasons for their predictions varied from an increase in libido or aggressiveness, to a general increase in their sense of well-being. On the negative side, testosterone produced a significant increase in the number of circulating red blood cells to a level which was at the upper limit of normal, and further increases would have increased the risk of polycythemia in which the blood becomes so thick that it does not flow properly. In addition, testosterone was also shown to stimulate (and therefore probably enlarge) tissue in the prostrate gland. This could be a serious problem since prostrate cancer is known to increase markedly in older men. Moreover, benign prostrate enlargement is very common and may be present in 60–70% of males who are aged 60 or over. Despite these possible drawbacks, Tenover points out that testosterone has been available for over 50 years and the cost of treatment is extremely cheap (about $100–$200 per year of treatment). Thus, there are reasonable grounds for using testosterone on a larger scale.

One of the most important medical uses for testosterone is in the treatment of sexual problems and, in particular, impotence. There is little doubt that sexual function in men declines with aging. For example, Alfred Kinsey in 1948 found that sexual activity as measured by the number of reported ejaculations per week decreased progressively from about 3 per week after the age of 30, to less than 0.5 per week by 70, and near zero by the age of 80. This decline in activity is also accompanied by a decline in sexual performance. In one study that compared a group of young men (aged 20–40) with a group of older men (aged 50–89), a number of differences between the groups were found. For example, the time needed to obtain an erection was 3–5 seconds in the young group compared with 10 seconds to several minutes in the old group; furthermore the distance the semen travelled followed ejaculation was some 12–24 inches in the young men compared to 3–5 inches in the older men. In addition, young men tend to have firmer erections and can more quickly re-engage in sexual activity than older men. It might be expected that testosterone has an important role to play in all these activities.

However, in practice, the relationship between testosterone and sexual performance is very complex. In general, research has not demonstrated a clear relationship between levels of testosterone and sexual performance in young or middle aged men. In older men the situation is less clear. It appears that low levels of testosterone may predispose the older person to experiencing sexual problems, although this is far from inevitable. Most researchers seem to agree that some minimum level of testosterone is needed to maintain male sexual activity and desire but, beyond this minimum level, the relationship between sexual activity and testosterone breaks down. In support of this view, testosterone administration for the treatment of impotence tends to be only effective in men whose testosterone levels have become extremely low or 'hypogonadal'. In fact, most cases of impotence are believed to be psychological in origin and not due to low levels of testosterone. Consequently its administration has little effect on improving the disorder. Thus, it appears that the decline in sexual function that occurs with aging in men cannot be solely blamed on testosterone.

Interestingly, there is some evidence that sexual activity increases testosterone levels and that this may be beneficial for health. For example, several studies have shown that rats involved in regular mating, or used for breeding purposes, show better condition and live longer. In one study, for example, the life span of male rats given an opportunity to mate at least once a week was 734 days compared to 578 days for unmated litter mates. The same experimenters also showed that the mated males contain 43% more testosterone than the unmated males. It is impossible to meaningfully relate these findings to humans although it is known that married people live longer than unmarried individuals, and that the average age of ceasing regular coitus has been placed at 68 in married men and 58 in those who are unmarried. Of course, nobody would seriously try to explain this mortality difference simply in terms of sexual inactivity or decline, although it is perhaps reasonable to assume that sex helps to increase the general sense of well-being, and is also probably beneficial to health.

In summing up, it is clear that there is a complex relationship between testosterone, sexual behaviour, well being, and aging. All of these different factors are inextricably linked – highlighting the great difficulty in trying to examine any single factor in isolation. It is undoubtedly naive to think testosterone holds the key to reversing aging and restoring sexual function; although low levels may be detrimental in middle aged and elderly males, and in these cases require treatment (although what constitutes a low level of testosterone remains far from clear). Even so, it is likely (although not certain) that the main benefits of such treatment would be increasing the sense of well being, rather than slowing down the rate of the aging process.

THE FEMALE REPRODUCTIVE DECLINE

Of course, men are not the only sex to undergo a decline in reproductive function. Although women tend to show a decrease in fertility throughout their reproductive years, the most conspicuous and dramatic age-related change is the menopause. This is not normally a sudden event but a gradual process lasting a number of years, in which the menstrual cycle becomes progressively less regular until it stops completely. In the majority of cases this event occurs between the ages of 45 and 55, and the end point of the process marks the woman's incapacity to reproduce. In contrast to most other species, many of which reproduce throughout their life, the human female's menopause occurs at a relatively early age in her life span. Indeed, most women can expect to live at least a third, if not half of their life, in a post-menopausal state. Despite this, the menopause is a significant event in any woman's life, not least because it heralds the start of what most people perceive as middle age.

The menstrual cycle is a remarkable interplay of complex hormonal events and body changes that normally takes place over a monthly period. In one sense, it is the body's way of preparing itself for pregnancy and, when this does not occur, the body discards what it has prepared and then starts the whole process again. The cycle begins (following menstruation) with the pituitary gland releasing minuscule amounts of follicle stimulating hormone (FSH) which then stimulates the development of a follicle (essentially a fluid sac containing an egg) in the ovary. As the follicle develops over the next two weeks it begins to release estrogen and, after about 14 days into the cycle, the oestrogen levels in the blood build-up sufficiently to trigger the pituitary gland into releasing another hormone known as luteinizing hormone

(LH). This hormone induces ovulation in which the egg leaves the confines of the follicle and enters the fallopian tubes where, if conditions permit, it can be fertilised. The follicle is not yet finished, however, and it continues to secrete oestrogen and another hormone progesterone ('pro-gestation') which results in the womb building up a lining ready for possible implantation of the fertilised egg. The egg has a relatively short period (approximately 3 days) in which to be fertilised and, if this does not occur, it quickly degenerates and dies. In turn, the lining of the womb, having no useful purpose, is discarded resulting in the menstrual flow. The average woman can be expected to experience some 350–400 menstrual cycles in her life time before the menopause arrives.

All this changes with the menopause. With the cessation of ovulation the ovaries begin to shrink and oestrogen and progesterone secretion markedly declines, so that within a few years their levels practically dwindle to zero. A small amount of oestrogen and progesterone may be produced by the adrenal glands and the breakdown of adrenal testosterone – but this typically makes up less than 5% of that which was present during the midcycle hormonal peak before the menopause, and of course, it does not follow a fluctuating monthly pattern. In reaction to the unresponsive ovaries, the circulating blood levels of FSH and LH significantly increase, in an apparently desperate attempt to stimulate the ovaries back into action, but to no avail.

The menopause is generally viewed as a normal stage of life and, in some cases, may even be regarded as a positive event, especially if the woman has previously suffered menstrual problems. Further, despite what is commonly believed, the vast majority of women experience no health problems, or psychological upheavals, following the menopause. Nevertheless, the menopause does have a number of important long term health implications for women. We have already noted, earlier in the chapter, the increased risk of heart disease when women lose the protective effects of their oestrogens on LDLs. But there are other protective effects of oestrogen that are unmasked following the menopause, and most worrying of all is osteoporosis, or loss of bone mass. On average, a woman will lose approximately 1% of her bone mass for every year she survives after the menopause. This means that if a woman has become menopausal at the age of 50, she will have lost a quarter of her bone mass by the age of 70, and half if she ever reaches 100. Not surprisingly, bone breakages become increasingly common after the menopause. Fractures of the wrist are the first to increase, with hip fractures becoming prevalent a few years later. By the time a woman reaches 80, a third will have suffered at least one major fracture (with potentially fatal complications occurring in about

15% of cases), and many of these victims will never again enjoy full mobility.

Oestrogen (and progesterone) also have some other important protective effects. For example, these sex hormones normally help sustain the fallopian tubes, uterus, vagina and breasts – and without their protection some degree of wasting typically occurs. Consequently, the uterus shrinks, the ovaries regress to about half their previous size, and the walls of the vagina (about 25 layers thick in young women) become thinner perhaps by as little as 2 layers in the very old. There is also an increase in some types of cancers, particularly breast cancer, with two-thirds of all breast cancers occurring in women over the age of 50. All things considered, therefore, the long term prognosis for women once they have undergone the menopause is far from rosy; although to put things into their correct perspective it must still be remembered that women still typically live longer than men. Nevertheless, recent evidence suggests that women who have an early onset of menopause (at ages less than 44 years) have a higher chance of dying earlier than those who undergo menopause at ages 50 to 54 years. Thus, the menopause is a potential health and aging risk of which every woman should be aware.

BIOLOGICAL CAUSES OF THE MENOPAUSE

Clearly the menopause is a major aging event and, therefore, it is important to understand what causes it to occur. In humans the answer appears to be that the ovary runs out of its quota of eggs. At first this may not appear to be too surprising until one realises that at birth the ovaries contain well over a million eggs, which is vastly more than the few hundred or so that will be released through ovulation during the women's reproductive life. In fact, for reasons that are not clear, most of these eggs disappear in the early years of life, with around three-quarters of a million eggs lost by puberty, and from then onwards the number of eggs in the ovary appears to halve every 7 years or so. Presumably, only the fittest eggs survive, although this is pure speculation.

Thus, the human findings appear to show that the main aging change of the female reproductive system is taking place in the ovaries. Surprisingly, this does not seem to be true for most other animals, who stop ovulating (and thus undergo their own menopause) whilst having plenty of eggs left in their ovaries. Therefore, in most species, the cause of reproductive decline appears to lie elsewhere in the body. This has further been demonstrated by research showing that ovaries taken from old mice and implanted into young hosts are generally able to start cycling again, whereas the same does not occur when young ovaries are placed into old mice. Therefore, in mice at least, old ovaries are still functional but, clearly, something in the aged body is stopping them from releasing their eggs (see also pages 141–142).

Why then does ovarian failure occur in animals when there are plenty of eggs left in the ovary? It appears that the answer lies in the brain or, more specifically, the hypothalamus and pituitary gland. In short, the evidence indicates that as the pituitary gland ages, it gradually releases less luteinizing hormone, until the amount of this hormone is so small that it is no longer able to trigger ovulation and initiate the period of oestrus (in humans this does not occur and luteinising hormones levels actually increase after the menopause). But why should luteinizing hormone release decline in this fashion? One might imagine that the answer lies with the hypothalamus or pituitary gland, although this appears not to be the case. As we have already seen, when ovaries are taken from young mice and grafted into old hosts, they stop cycling despite plenty of eggs being left. However, there is an important exception, and this occurs when young ovaries are implanted into old animals that have had their ovaries removed at an early age. Surprisingly, the transplantation of young ovaries back into old and 'ovary-less' animals is often able to kick-start their oestrus cycles back into action.

How can we explain this puzzling finding? The main difference between the two groups of recipient animals is that one group (those with intact ovaries) have produced oestrogen throughout their life, whereas the other group (those without ovaries) have not produced oestrogen. And the obvious conclusion is that the oestrogen must have 'aged' the hypothalamus and pituitary gland, and reduced their ability to secrete luteinizing hormone. This theory has been supported by Caleb Finch and his colleagues at the University of Southern California who have shown that, under normal conditions, laboratory female mice undergo around 50 oestrus cycles during their life; but this figure is significantly reduced if mice are given high amounts of oestrogen beginning early in life. That is, oestrogen exposure appears to speed up their reproductive decline.

Rather than being beneficial, oestrogen may therefore have significant aging effects. In fact, Finch goes one step further and argues that oestrogen acts to shut down the oestrus cycle by causing cumulative and irreversible damage to the hypothalamus. Various calculations have led him to propose that once the animal has been exposed to 5,000 units of hormone (equivalent to 50 oestrus cycles) then sufficient damage will have occurred to the hypothalamus (and possibly pituitary gland) to stop the release of luteinizing hormone, and thus ovulation. If Finch is right, then it can be seen that every oestrus cycle with its concomitant increase in oestrogen secretion is, in effect, acting like an aging clock, and is gradually timing the countdown of reproductive decline. That is, each oestrogen surge is a tick towards the final reproductive breakdown. Human females have apparently escaped from this type of 'clockwork' control since they use up all their eggs before reproductive decline sets in. One can only speculate as to the reasons why evolution has allowed this situation to occur in humans.

Despite this, there is some evidence to indicate that oestrogen might be able to produce neural change in the human hypothalamus. This evidence comes from hot flushes (which affect up to 75 to 80% of menopausal women) that result from oestrogen deficiency. Hot flushes are due to sudden bursts of capillary expansion that produces a warm burning sensation, often located around the head or the neck, which can also be followed by increased perspiration all over the body. Although the cause of hot flushes are not fully understood, current evidence suggests that they are due to a malfunction of temperature control mechanisms in the hypothalamus. Furthermore, it appears that pre-exposure to oestrogens in puberty is a prerequisite for them to occur. This can be shown by the fact that females with Turner's syndrome (which is a disorder where the ovaries do not develop) only experience hot flushes if they have previously been given oestrogen treatment. If they have not been given oestrogen as a treatment for their disorder then hot flushes do not occur. It appears, therefore, that early exposure to oestrogen may prime the nervous system in some way and the most probable site for this effect is the hypothalamus.

THE USE OF HORMONE REPLACEMENT THERAPY

Although there is a great deal of controversy surrounding the risks and benefits of hormone (or more specifically oestrogen) replacement therapy (HRT) in treating menopausal symptoms, there is nevertheless strong medical evidence that it can alleviate many of its problems. Perhaps most convincing are the findings showing that HRT has a protective effect against osteoporosis. For example, Roger Gosden in his book *Sex and Aging* has written that osteoporosis is almost stopped in its tracks by HRT treatment, and that prolonged administration can reduce bone fracture rate by as much as 70%. Indeed, few doctors now doubt the usefulness of HRT for this condition, although it is unclear to what extent the benefits are maintained after the therapy is terminated. Perhaps less widely known is the protective effect of HRT on heart disease. In fact, more than 30 studies have shown that postmenopausal women who use oestrogen are at lower risk for coronary disease than those who do not use oestrogen replacement. This is particularly strong justification for using HRT since coronary heart disease is the main cause of death for women in Western society, killing more females than all other forms of cancer combined. It has been found that oestrogen replacement in the first 10 years after menopause may reduce mortality from heart disease by as much as 50%, and that the benefits are even more striking in women who already suffer from coronary heart problems. It is also well established that the incidence of heart disease in women increases considerably after the menopause, and that the risk of heart disease is greater in those who undergo an early menopause than those who go through it later.

Despite these clear-cut benefits there are potential risks with the use of HRT. For example, oestrogen increases the chances of developing cancer of the uterus (although adding progesterone to the treatment may reduce this risk) and there also appears to be a small but significant risk of breast cancer. Nevertheless, it is probably fair to say that, for most women, the benefits of HRT appear to outweigh the disadvantages, at least in the short term. This is supported by findings that have examined mortality in women who have used HRT. For example, in a study published in 1991 by Brian Henderson at the University of Southern California, who followed 8,881 post menopausal women for a period of over 7 years, it was found that women with a history of oestrogen use enjoyed a 20% reduction in mortality *from all causes.* Furthermore, this figure rose to 40% in women who were current users and who had been using oestrogen for at least 15 years. However, it must be remembered that hormone replacement is not a permanent treatment, partly because women tolerate it less well as they get older; and evidence suggests that death rates, particulaly from heart disease, revert back to normal within a few years of HRT cessation. Thus, it seems unlikely that HRT significantly increases long term or maximum life span, although it would seem to protect against certain aspects of aging and premature death.

Despite these positive findings, critics have pointed out that in all of the studies that have so far examined HRT, the women who have been given hormone replacement have themselves chosen, or sought therapy, and are likely to differ from women who have not opted for this treatment. For example, it is probable that women seeking HRT will be from a higher socio-economic class, be more health conscious, take more exercise and be more concerned with weight control. Indeed, women who are placed on HRT will see a doctor regularly and are, therefore, more likely to have health check-ups (including those for blood pressure and cholesterol levels). Consequently, it is conceivable that the reported gains in health and life expectancy following HRT may be biased, since such women are healthier in the first place. Sadly, it is difficult to see how this particular problem can ever be success-fully overcome without a large scale study covering all social groups and a large degree of subterfuge. Thus, a question mark is always likely to hang over the true usefulness of HRT, although from an indi-vidual's perspective it is clear that the vast majority of women perceive the effects of hormone replacement to be positive. For this reason alone it should not be under-valued.

SUMMARY

• In 1849, Arnold Berthold performed the first experiments showing that the testes were secreting a chemical that was determining the sexual characteristics of male roosters.
• In 1889, Charles Brown-Sequard claimed that testicular extracts taken from a number of animals were able to rejuvenate declining sexual function and reverse aging. Despite the great media interest his idea received at the time, it is generally believed that Brown-Sequard was mistaken.
• There have been many other scientists who have followed in Brown-Sequard's footsteps and claimed that medical procedures to reju-venate the sexual organs can have anti-aging effects. These include Eugen Steinach, Sergi Voronoff, and Paul Niehans.
• The male sex hormone produced by the Leydig cells of the testes is testosterone. It is released in response to luteinizing hormone which in turn, is produced by the anterior pituitary gland.
• Most studies have tended to show that circulating levels of testos-terone decline with age although this is not necessarily an inevitable feature of aging. This decline is probably due, in part, to a reduction in the number of Leydig cells.
• In terms of world records: regular coitus with mutual orgasm has been recorded between a man of 103 and his wife aged 90, whilst the oldest age at which a man has fathered a child is recorded at 94 years.

- It is difficult to see how testosterone can be beneficial for longevity. For example, it tends to increase levels of low-density lipoproteins (which have been linked to heart disease) and, furthermore, eunuchs appear to have respectable life spans. Also, on average, women tend to outlive men.
- Contrary to popular belief, testosterone is not usually a successful treatment for impotence. There is a complex relationship between sexual behaviour, testosterone, well being and aging.
- Women undergo reproductive decline in the form of the menopause which results in the cessation of ovulation with consequent loss of monthly patterns of oestrogen and progesterone. It appears to be largely due to the ovaries running out of eggs.
- It is clear that in some women there may be long term health risks following the menopause including increased risk of osteoporosis and heart disease. The use of hormone replacement therapy would appear to reduce, or delay, the post-menopausal onset of many health problems, although the issue still remains controversial.

THE IMMUNOLOGICAL
BASIS OF AGING

*The theory of aging I love best and am inclined to be more
faithful to than others is the immunological theory, which posits
a double dose of doom arriving with the years, one part decline
of function, the other part active destruction*

–Roy Walford

*The vulnerability of old age depends on the progressive
inadequacy of the immune system.*

–Macfarlane Burnet

It is easy to forget that our bodies are a constant battleground alive
with the ebb and flow of war. Throughout our lives we are continually
being bombarded by micro-organisms that find our warm moist bodies
an agreeable place to live and reproduce, and without an immune
system they would quickly overwhelm us. These invaders include
viruses, bacteria, moulds, yeast and protozoa – all of which can cause
serious illness and death if left free to colonise our tissues. Thus, to
remain alive and remain in good health, our immune system has to be
constantly on guard to protect us from such marauding attackers. But
the immune system does much more than simply protect us from these
external threats. For example, it defends us from our own cells, if and
when they start to become abnormal (as may happen in the initial
stages of cancer), and it also helps in clearing away unwanted cellular
debris that can accumulate in our tissues. The immune system is a particularly complex and intricate system, and one that not unexpectedly
shows a decline with aging. However, this decline has many more ramifications than first meets the eye. Obviously, we might expect to become
less resistant to infection, and perhaps more susceptible to cancer with
aging – but it is also clear that a declining immune system might also
become impaired in its ability to distinguish *self* from *nonself*, and in
consequence start to attack its own body tissues (autoimmunity). In
fact, some theorists not only believe that autoimmunity occurs with
aging, but that it may actually be the main cause of aging – that is, we

age because our immune system turns against us. Indeed, like the hormonal system, there is no part of the body which is not susceptible to the effects of a declining or wayward immune system and, as we shall see, there are many reasons to suspect its involvement in aging.

INTRODUCTION TO THE IMMUNE SYSTEM

Before looking at how the immune system ages, we first have to familiarise ourselves with the system itself. For the unacquainted this is no easy task since the immune system, and the way it works, is a particularly complex subject. A good place to begin is with the various structures of the immune system that are distributed throughout the body. These include the red bone marrow (in particular the marrow found in the ribs, vertebrae and pelvis), thymus gland, spleen, and lymph nodes. All of these organs serve many functions; but essentially they produce a variety of specialised cells that protect the body against attack, and which become involved in search and destroy operations once the body's defences become breached. These cells are to be found either stored and primed for action in the organs just mentioned, or circulating in the lymph and blood systems which allow them access to every site and cell of the body.

If one is able to take a drop of blood, add the appropriate dyes, and then examine the smear under a microscope, a number of cells can be seen. By far the most common type are the red blood cells that give our blood its colour and which are involved in transporting oxygen from our lungs to our tissues. In fact, these are so numerous that they will make up over 99% of the cells on our glass slide. The remaining cells are the so-called white blood cells which are essentially the patrolling soldiers of the immune system. There are at least five types of white blood cell (neutrophils, eosinophils, basophils, monocytes and lymphocytes) that are otherwise known as leucocytes (*leuco* means 'white' or 'lacking in colour'). We will discuss some of these cells in more detail later, but suffice to say, they each have their own specialised function in immunological defence.

A remarkable feature of our blood cells (both red and white) is that they are all derived from one single type of cell located in the bone marrow. This unique cell is called a stem cell (see also page 152) and it is basically a primitive cell that has no function other than to replicate and, in the process, create new and more specialised cells. Stem cells are effectively immortal because when they divide, they make one identical copy of themselves (a 'clone'), along with a new type of cell that is slightly more complex. It is these later cells that go on to divide further, and to eventually develop (or 'differentiate') into the many different kinds of blood cell. Much of this differentiation goes on in the bone marrow, although some cells, notably the monocytes and lymphocytes, leave the bone marrow in an immature state and become immunocompetent elsewhere in the body. As might be expected, the bone marrow is in a constant state of cell division. It has been estimated that it produces 2.4 million new red blood cells *every second*, as well as producing approximately 10 million new white blood cells each day. These numbers increase further during times of infection.

One of the key secrets to understanding the immune system, however, is an appreciation of the lymphatic system. When we think of the body's organ systems the lymph system rarely springs to mind, yet without it our cardiovascular system would stop working and the effectiveness of our immune system become fatally weakened. The need for a lymphatic system arises because of a major plumbing problem that occurs as the blood delivers oxygen and nutrients to the watery and nutrient rich fluid bathing our cells. The high pressure in the minute and narrow blood capillaries means that there will always be some degree of seepage of plasma constituents before the nutrients reach their destination. Consequently, the lymph system acts as a drainage mechanism collecting all the escaped fluids and eventually returning them back into the circulation via the great veins of the neck. However, mammals have also adapted this system to serve the needs of the immune system. Not only does it provide an important transport system for immune cells, but it has also evolved 'inspector stations' or lymph nodes throughout its intricate network that are able to examine and filter the constituents of the lymph in minute detail. These nodes contain a variety of important cells including macrophages (that 'eat' foreign material) and other types of immune cell called lymphocytes (see below). Because the lymphatic system drains the excess fluid from every small part of our tissues, it provides an invaluable means of detecting the presence of foreign invaders that have managed to slip into our bodies. Indeed, rarely are they able to avoid being sucked into the lymph system where they set-off the immune system's alarm bells.

NATURAL IMMUNITY

Our first line of protection is the body's own innate defence (i.e., the immune system we are born with). This system contains many components including physical barriers (such as the skin and mucus membranes that line our respiratory and digestive tracts); phagocytes (cells that are able to recognise, engulf and break down foreign cells and debris); and various blood-borne molecules. To get an idea of how this system works, let us imagine we fall over and graze our leg. Our injury is likely to initiate two basic reactions: firstly tissue damage will set up swelling and inflammation around the wound thereby attracting a number of immune cells to the area; and secondly, any foreign material entering our body will be swept up into the lymph fluid and carried into the lymph network where it will come across a lymph node. These nodes can be likened to sentry posts where everything in the lymph is dissected, probed and analysed. They also contain large numbers of phagocyte cells, the most important of which are macrophages (meaning 'big eater') that have prodigious appetites and consume (in a process called 'phagocytosis') any foreign material or cell debris that they meet. In fact, during the first hour or so after infection, these macrophages provide us with our first crucial line of immune defence.

However, other processes are also being put in motion whilst the macrophages are at work and, in particular, the inflammation around the initial wound will be attracting large numbers of a second kind of white blood cell called neutrophils. These make up our most abundant type of white blood cell (amounting to about 60% of all leucocytes) and are scavenger cells (like macrophages) that are intent upon consuming any foreign particles they encounter. The neutrophils are small and very mobile, and are able to pass out of the blood vessels into the surrounding tissue (often helped by the process of swelling) to continue the search for infecting pathogens. This response may take a little while to build up, but within a few hours after the onset of acute inflammation, the number of neutrophils in the blood may increase four or even five-fold, reaching 25,000 per cubic millimetre. Thus, within several hours after tissue damage, the area is likely to be awash with neutrophils seeking out foreign antigens.

This is not the end of the natural response, however, since there is yet another defence system ready to lend assistance, and this involves the large scale migration of monocytes into the damaged area. Monocytes are immature white blood cells, but as they enter the wounded tissue they transform themselves into macrophages, and as with the macrophages in the lymph nodes, these transformed monocyte cells vociferously consume everything in their vicinity. These new cells can destroy far more bacteria and consume much larger particles than the neutrophils; and may even consume the neutrophils if they are showing signs of degeneration after eating their fill of foreign material. This third line of defence may take around 8 to 12 hours to begin, but once set in motion it provides a much sterner defence against the infection.

ACQUIRED IMMUNITY

One problem with innate immunity is that it cannot provide protection against all possible infections that an organism is likely to meet. For example, microbes evolve rapidly in a changing world and, in doing so, often devise ingenious ways to evade the body's defences. This not only poses a threat to the individual but also poses a real risk to the species as a whole. Furthermore, innate immunity is rather slow, and some infectious agents can do considerable damage before they are overcome. The possibility of repeated infections by virulent agents provides yet another threat. Because of these dangers, vertebrates have evolved a second type of immune response that is known as acquired immunity in which antibodies have an important role. This form of immunity not only enables the body to adapt and to respond to any microbe that has escaped the clutches of the innate system (even if it has never come across the pathogen before); but it also enables the immune system to 'remember' the infectious agent allowing it to be primed for action should it be encountered again. Thus, the immune system is effectively able to 'acquire' a response that was lacking previously. Although acquired immunity probably arose originally because our innate defences were not foolproof, it has now evolved so that it works in conjunction with our innate immune system. In one sense, therefore, it is meaningless to talk of innate and acquired immunity separately because they work together.

Acquired immunity depends upon a different class of white blood cell known as lymphocytes. Although these make up about 20% of all white blood cells, only a small proportion of their total number are found in the blood and, as their name suggests, most are to be found located in lymphatic system including the lymph nodes (where they form large colonies), spleen, tonsils and the thymus. There are 2 main groups of lymphocytes; the T-lymphocytes or T-cells (so called because they have to pass through the thymus before they become functional) which account for around 80% of the lymphocytes found in the blood: and the B-lymphocytes or B-cells which are involved in the formation of antibodies (the 'B' derives from the *bursa of Fabricus*, a small patch of lymphoid tissue found in the intestine of birds where these lymphocytes were first discovered). These two types of lymphocyte have quite different functions and, as we shall see below, are involved in two separate forms of acquired immunity.

One of the earliest references to acquired immunity can be traced back to 430 BC when the Greek historian Thucydides wrote that only those who had survived the plaque should be allowed to nurse further plaque victims as they would be immune to the disease. A similar situation arose in the middle ages with the horribly disfiguring disease small-pox. As with the plaque, only those who were fortunate enough to sur-vive the illness seemed to be protected from further infection. Although at the time people did not understand what caused the disease, some had nevertheless discovered that resistance to smallpox could be obtained by rubbing the powered scabs from smallpox victims into the grazed skin of healthy individuals. Unfortunately, the treatment did not always work, and some patients developed the full-blown disease and died as a consequence. In fact, it was not until Edward Jenner in 1718, that a safe form of inoculation to smallpox was found. Jenner was intrigued by the fact that milkmaids who caught cowpox (a much milder disease) never contracted smallpox, and he hit upon the idea of protecting people from smallpox by exposing them first to cowpox. He tested his theory by scratching into the skin of an eight-year-old boy the pus taken from a cowpox sore, and then less than 7 weeks later exposing the same child to smallpox. Fortunately for the child (and Jenner) the procedure worked and the child was found to be protected against small-pox. This new technique become known as vaccination (after the Latin *vacca* meaning cow) and quickly become accepted medical practice.

The main reason why Jenner's procedure worked is because of a process called humoral immunity that involves the formation of anti-bodies that travel in the body's 'humours' or fluids. Antibodies are pro-tein molecules that are formed in response to an infection, and which are tailor-made to recognise the invading infectious agent (or 'anti-gen'). Unfortunately, the development of antibodies takes the immune system several days, and in the case of severe infections such as small-pox, their production may be too late to be of any help. What Jenner had unwittingly done was to make the immune system develop anti-bodies to the cowpox virus, that in turn had managed to provide pro-tection against smallpox. The reason why this procedure worked was because the two viruses are sufficiently similar (they are in fact more than 95% identical) that the antibodies formed against cowpox also reacted against the more dangerous smallpox virus.

We now know that the formation of antibodies begins with lympho-cytes (B-cells) that tend to be found in the lymph nodes. There are huge numbers of B-cells all of which are slightly different, and each one has studded in its outer coat an antibody molecule that is designed to fit just one foreign antigen molecule. Should a B-cell find an antigen that fits its own individual antibody receptor (much like a lock and key), it will then start to replicate itself. Most of these new B-cells will then enlarge and mature into what are known as plasma cells. These are basically cellular factories for making huge numbers of their own individual antibody proteins. It takes several days for a plasma cell to develop, but once they are fully formed the amount of antibodies they produce is astounding – around 2,000 molecules per second! Some of these antibodies remain attached to B-cells, whereas others drift free and are released into the lymph and blood, where they roam the body seeking out the antigen that they are designed to recognise.

This is a rather simplified account since the B-cell also requires an extra prompting to start its transformation into an antibody secreting cell. As we have seen above, macrophages (or so called 'big eaters') are also found in the lymph nodes where they consume almost everything that comes their way. But rather than simply digest the contents of what they 'eat', macrophages also regurgitate parts of their victims; and monitoring the macrophages for this type of event is a special type of T-cell known as a 'helper cell'. When these cells come across a macrophage fragment which they do not recognise, they respond by dividing and releasing special chemicals called lymphokines. These in turn, stimulate the B-cells, and with this extra confirmation the cell then goes ahead with the production of antibodies.

Antibodies are simply Y-shaped proteins whose two short arms latch onto their antigen with great specificity, and once firmly in place, set into motion a number of immune processes leading to the antigen's destruction. Although antibodies can cause some damage to infectious agents directly, they are relatively inefficient in this task and, consequently, their main function is to 'flag' the invader which then alerts other cells of the immune system to take over the function of killing. Indeed, one of the main functions of antibodies is to orchestrate and amplify the basic innate immune responses. When the antigen's destruction has been accomplished, and all the remains of the antigen cleared away, the production of antibodies then stops. However, the immune system still retains large numbers of sensitised B-cells, and these remain functional in the body for long periods of time. They are used as 'memory cells' in order to provide a faster, prolonged, and more efficient response should the antigen ever be encountered again.

CELL-MEDIATED IMMUNITY

The basic essentials of the antibody response had been worked out by the end of the 19th Century. However, it was not until the 1950s that immunologists began to realise that there might be another type of acquired immune response that did not involve antibodies. This form of immunity is now known as cell-mediated immunity and it utilises several types of T-lymphocyte as its main weapon. In some respects, T-cells are similar to B-cells since they contain molecules in their outer coat that are like antibodies (called surface receptor proteins) which are specific for one particular antigen. Furthermore, once T-cells are activated they quickly reproduce and, in the process, create large numbers of themselves. T-cells do not release their receptors into the circulation as antibodies, but they head off (armed with their surface receptor proteins) looking to defend the body against invaders.

The secret of understanding how T-cells work lies in the way they recognise foreign agents. T-cells are unable to recognise invaders freely roaming in the lymph or blood (unlike B-cells), but can only detect antigens that are attached to the surfaces of the body's own cells. This makes T-cells particularly important for mounting attacks against the cells of the body that have become infected by viruses and bacteria, or become abnormal or cancerous. However, to recognise an antigen presented in this way, a T-cell must be able to distinguish what is *self* from *non-self*. For example, since our body cells have many normal proteins in their outer membrane coat, how do T-cells know which ones are to be attacked? The answer is that all of our cells have on their surface a special set of identity proteins (called MHC proteins) that T-cells recognise as 'self'. Everyone of us (except for identical twins) have a slightly different set of these self markers, and this is one reason why tissue transplantation between different individuals generally fails (in this case the T-cells do not recognise the new MHC proteins as 'self' and therefore attacks the 'foreign' tissue).

In fact, our MHC proteins have a dual role because they are also the sites where foreign antigens are presented to the T-cells. Similar to the macrophages that regurgitate parts of their victims to B-cells (see above), our normal body cells also present foreign proteins on their surface, and they do this by attaching them to the MHC proteins. When the happens the shape of the MHC-fragment complex becomes changed and is no longer recognised by the T-cell. In response, the T-cell starts to change, and one of the ways it does this is to transform into a cytotoxic or 'killer' T-cell. These cells not only attack the crippled body cell, but they also travel throughout the body, circulating in and out of the blood, and lymph, in search of further body cells displaying antigens to which they have become sensitised. It is not clear how cyto-toxic T-cells kill infected cells, but there is some evidence that they may induce apoptosis (see pages 95–97) or they inject into the cells a chemical called perforin which causes its death by making the mem-brane 'leaky'. Whatever the mechanism, it is important to remember that these cells attack the body's own cells if they have become infected or damaged.

Cytotoxic T-cells are not the only type of T-cell to be activated follow-ing infection. We have already come across helper T-cells, but there are others including suppresser T-cells (whose main function is to dampen the activity of the immune system) and memory T-cells (that become primed following an infection). In fact, these T-cells are actually stimu-lated by a different type of MHC protein although this not need concern us here.

Table 10.1 A brief summary outlining the characteristics of the main divisions of the immune system.

	Natural Immunity	Acquired Immunity	
		Antibody	Cell-mediated
Main Type of Cell	Phagocytes	B-Cells	T-cells
Main Location	Macrophages are found mainly in the lymph nodes. Neutrophils are also found in the blood	B-cells are found circulating in the blood and lymph. They are also stored in the lymph nodes	T-cells are found circulating in the blood and lymph. They are also stored in the lymph nodes
Main Function	Engulf foreign particles (phagocytosis)	The production of antibodies, also including memory cells	A number of roles including killing antigen-bearing cells of the body
Site of Origin	Bone marrow	Bone marrow	Bone marrow, but then T-cells migrate to the thymus where they become immunocompetent

THE IMMUNE SYSTEM AND AGING

As with most other biological systems of the body, it is clear that the efficiency of the immune system declines with age. Indeed, common sense tells us that such a process is at work. For example, a child is typically able to shake off a bout of influenza after a day or two, whereas the same illness will nearly always cause a more severe and longer bout of illness in an adult. Further evidence of this relationship comes from mortality data that follows influenza epidemics. Following an outbreak of influenza, nearly every death that results from the illness will occur in those over the age of 60 and, in particular, those who are already infirm or ill. There is also evidence that infectious diseases are a leading cause of death amongst the very old. For example, Robert Kohn in a study that looked at the causes of death among 200 people over the age of 85 years, found that infections (that were not secondary to another main cause of death) accounted for 17% of all deaths, with the most common infections being pneumonia (making up over 50% of all deaths) and influenza. Thus, by old age it seems that the immune system is beginning to lose much of its protective function. Why then does such a decline occur, and what components of the immune system are most affected?

AGE CHANGES IN THE THYMUS GLAND

Undoubtedly, the most striking age-related change of the immune system occurs in the thymus gland, a pinkish greyish mass situated just below the neck in the chest (the thymus was given its name by Galen in the second century AD because of its resemblance to a bunch of thyme leaves). The thymus grows to its full size in late childhood reaching a weight of approximately 30 g, but then it dramatically begins to shrink, and this continues unabated throughout adolescence and adulthood so that by the age of 50 years it only retains 5–10% of its original mass. Moreover, by the age of 70 it has almost disappeared although it does not normally disappear completely. The reason why the thymus declines in this way is not known, although it might be that it has performed its most important work by adolescence, and consequently has relatively little work to perform in adulthood. Despite this, many researchers believe that the involution of the thymus gland is also an important factor in the decline of the immune system, and that it contributes to our increasing susceptibility to disease.

What then is the function of the thymus gland? Until the 1960s the answer to this question was not known since removal of the thymus gland in adult laboratory animals produced no obvious change. Moreover, in humans, its inaccessible location in the top of the chest meant that its function could not be measured. It appeared to be redundant. However, this view changed with the work of Jaques Miller in 1961 who showed that immune deficits could be demonstrated if the thymus was removed in new-born mice. In short, Miller showed that these mice (unlike normal animals) did not reject skin grafts from unrelated strains of mice. Since graft rejection is known to be dependent upon the ability of T-cells to recognise foreign tissue, this suggested that the thymus gland was somehow involved in helping the immune recognition process, and that it conferred this function on T-cells early in life.

Indeed, it is now well established that the thymus is a site where T-cells mature. As mentioned above, T-cells actually begin their life in the bone marrow, but then leave as immature lymphocytes and migrate to the thymus where they fully develop. During their maturation in the thymus two important things happen. Firstly, the lymphocytes either become helper T-cells or cytotoxic T-cells: secondly these cells acquire the ability to recognise self from non-self. This, of course, is crucially important since if T-cells were unable to recognise the cells belonging to their own body, they would attack all cells, both self and non-self, indiscriminately. In fact, it appears that the thymus takes great care to ensure that T-cells are able to perform this discrimination with a very high degree of accuracy since it takes some 2–3 weeks to process a T-cell and, even then, only about 5% of lymphocytes entering the thymus are released fully mature into the blood where they travel off to colonise the lymph nodes and other types of lymphoid tissue.

The thymus does most of its work early in life and it would appear to be stockpiling the various structures of the immune system with fully matured T-cells. Following childhood the thymus winds down its activity, perhaps just enough to keep levels of T-cells ticking-over. Indeed, thymic activity still continues in the old, and this may be one of the reasons why, despite its degeneration, there is no clear evidence of a decline in the number of circulating T-cells with aging. Nevertheless, there is experimental evidence that the thymus gland does become functionally impaired with aging. For example, in studies where mice have had their thymus glands removed at an early age, T-cell function was restored by the transplantation of thymuses from young, but not old, animals. Findings from human studies also supports this idea. For example, it has been shown that 85% of the lymphocytes taken from the thymus of a 20 year old person are 'matured' and capable of immune responses, whereas this number drops to 50% in cells taken from an 80 year old.

The thymus also produces a number of hormones (called thymosisns) that have an important role to play in the development of T-cells in the thymus, as well as helping maintain T-cells located elsewhere in the body. The levels of these hormones also decline with aging, and this is so marked that by the age of 60 they cannot normally be detected in the body. Again, this is further evidence that the thymus not only declines with age, but also becomes seriously impaired in its immune responsibilities.

AGE CHANGES IN T-CELLS

A classic test of cell-mediated (T-cell) function is the delayed hypersensitivity test that has become standard practice in hospitals as a means of testing skin allergies. An illustration of this technique is perhaps best exemplified by the tuberculin test. If an individual has come into previous contact with the tubercle bacillus they will have become sensitised to some of its own unique proteins. To test for the presence of these sensitised immune cells, a small amount of prepared tuberculosis antigen can be injected just under the skin and, if the individual's T-cells are indeed sensitised to this antigen, a visible and inflamed lump will appear after about 24 hours. An immediate response will show the presence of antibodies (or B-cells) that are sensitised to the tubercle antigen, but in practice most individuals show a much stronger cell-mediated response to this antigen which takes longer to mount. In fact, this reaction is caused by the activity of helper T-cells that respond to the antigen by releasing chemicals known as lymphokines, which in turn attracts other immune cells (notably macrophages and cytotoxic T-cells) to the area.

If T-cell function does indeed decline with age then we would clearly expect this delayed response to be reduced. In fact, this appears to be the case. For example, in one experiment performed by Ian Roberts-Thomson and his colleagues at the Royal Melbourne Hospital in Australia, delayed hypersensitivity reactions to a number of commonly used skin antigens were compared in young subjects under 25 years of age, and those who were over the age of 60. The results showed that all of the young subjects reacted to at least two of the tested antigens, whereas less than half of the old group demonstrated the same response. In fact, nearly 25% of the old subjects showed no response at all to any of the antigens suggesting a clearly impaired T-cell function. But, this was not all the researchers found: in addition to testing for hypersensitivity they also followed their subjects over the next two years and found that the very old subjects (over 80) who had poor responses to the antigens, also had greater mortality than that of comparable people showing strong responses. In other words, the results showed that a vigorous T-cell response is a very good predictor of health and longevity.

T-cells are also the main type of immune cell involved in the rejection of organs and tissues following transplantation, and we might therefore expect older individuals to be more able to accept foreign grafts. This idea has been supported in a number of animal studies and, indeed, it is generally accepted that the rejection of grafts is impaired (less likely) in the elderly. But the decline in T-cell immunity might have more far reaching consequences. For example, some researchers believe that T-cells are involved in the immune surveillance of the body's own tissues, and destroy any cells that become abnormal. The obvious abnormality that can develop is cancer. In fact, it has been estimated by one expert that we may develop at least one cancerous cell every day, but normally these are destroyed as fast as they arise. If this is the case, and if T-cells are involved in helping this process, then clearly, the decline of T-cell function may have more serious implications.

AGE CHANGES IN B-CELLS

There does not appear to be any significant decline in the number of circulating B-cells, or levels of antibodies with aging. The question of whether there is a decline in their ability to mount an immune response, however, is less easy to answer. As discussed above, it is important to remember that their are two types of antibody response: primary and secondary. When the immune system is first exposed to an infectious agent, it stimulates a primary response causing new antibodies to appear in the blood after 5 to 7 days, and these continue to proliferate so that they reach a peak after about 2 weeks. Once the antibodies have done their job, however, their levels return to a low resting level. Should the immune system be re-exposed to the antigen at a later date, then a secondary response is initiated. In this event, the antibodies are now primed to recognise the antigen, and consequently they are able to mount a much more efficient and rapid response.

In general, the evidence appears to suggest that while the primary antibody response declines with age, the secondary response remains unaffected. For example, several studies have shown that B-cells taken from aged individuals produce fewer numbers of antibodies when reacting to an antigen for the first time. In contrast, there does not appear to be a difference between young and old in terms of antibody production following a secondary immune response (this may also explain why childhood diseases are rarely found in old adults). However, despite this, researchers have long been aware that the primary antibody response also requires the assistance of helper T-cells, and there is increasing evidence that it is actually the functional decline of this T-cell that is the main culprit in the demise of the primary immune response. Thus, it is possible that no age-related decline occurs in B-cell function – which is in stark contrast to the T-cells where such decline certainly occurs. Despite this, there is considerable uncertainty over this conclusion, and it may be more accurate to say that if a decline occurs in B-cell function, then it is certainly not as great as that which occurs with T-cells.

THE IMMUNOLOGICAL THEORY OF AGING

One of the main implications of an aging immune system is that we will become much more susceptible to disease and infection as we get older. But this is not the only drawback of declining immune function. In order to defend against foreign antigens, our immune system must be able to distinguish the cells of the body from those of the attacker, and as we have already seen, the thymus gland goes to great lengths to ensure that its T-cells can perform this discriminatory function to a high degree of accuracy. But, of course, a declining immune system may also mean that this discrimination becomes increasingly impaired with the consequence that it will be more likely to turn upon its own tissues and inflict considerable damage (autoimmunity). Taking this idea one step further it is easy to imagine that such a process may also be an important cause of biological deterioration that we recognise as aging.

The possibility that autoimmunity might be linked with aging was first suggested by the Australian Nobel Prize winning immunologist Macfarlane Burnet in 1959, and later by Roy Walford in 1962 (although neither developed their theories fully until much later). Both theories overlap to a large extent and are generally in agreement that aging is largely a result of the immune system turning against its own body. Thus, when most people refer to the immunological theory of aging they are generally talking about both theories. However, this can be confusing since both theories also tend to emphasise different aspects of the self-destruction process, and in some regards are quite different. This makes the immunological theory one of the more complex aging theories discussed in this book.

In Burnet's view, the main reason why the immune system declines (and autoimmunity occurs) is because of a gradual decline in the discrimination of T-cells to mount proper immune responses, aided by the degeneration of the thymus gland. It is important to remember that a T-cell begins its life as a stem cell (in the bone marrow) and then has to make its way to the thymus where it is given the code to recognise the body's own cells. When the cells from the bone marrow reach the thymus they proliferate (divide) at a very fast rate, and in the process, vast numbers of cells are rejected (presumably because they do not meet the high 'self-recognition' standards required by the immune system). In other words, a lot of cell generations are used up in the thymus to provide relatively few mature T-cells.

According to Burnet this process helps explain why autoimmune reactions are likely to increase with aging. As we saw in chapter 6, the dividing cells of the body can only divide a certain amount of times before they reach their Hayflick limit – and in Burnet's view, since T-cells proliferate so rapidly, they become the first type of cell in the body to reach this limit. At around the same time, the thymus gland also begins to shrink. The reasons for this are not clear. It may partly be in response to the declining numbers of lymphocytes, although this is probably not the only answer (in fact, Burnet calls the thymus the *pacemaker* of aging which implies a cause of atrophy *intrinsic* to the thymus). Whatever the reason, the thymus becomes impaired in the processing of the now diminishing numbers of T-cells and, as this occurs, the chances of autoimmune (T-cell) dysfunction increases. This theory thus regards immunological decline (and therefore aging) as something that is *programed* to occur because of the way T-cells are limited in their ability to divide, and because of the clock-like shrinkage of the thymus.

However, Burnet adds a further strand to his theory by pointing out that all the cells of the body are also constantly accumulating different types of damage as part of day-to-day living. And of particular importance in this respect are somatic mutations (see page 84). Although somatic mutations can occur in all cells, according to Burnet, they are especially problematical if they arise in the cells of the immune system. For example, if a somatic mutation occurs in a mature T-cell, not only would such a cell be more likely to attack the body (i.e., its discrimination of self and non-self be impaired), but perhaps more importantly, if it ever formed part of an immune response (which required its proliferation) all of its cloned descendants would also be likely to carry the somatic mutation. In a young organism this may not be a problem since the rest of the immune system would probably destroy the mutant T-cell before it could give rise to any more descendants. But should the immunological system decline, then the mutant T-cells may well escape this censorship and go on to do considerable damage. It is also important to remember that T-cells play many other integrative roles in the immune system (e.g., helper T-cells are involved in the activation of B-cells and antibody formation.) Bearing this in mind, Burnet also speculated that mutated T-cells might also cause the B-cells to make antibodies against its own tissues (autoantibodies). Thus, the decline of T-cell function may have ramifications throughout the immune system.

A similar type of theory has also been proposed by Roy Walford in his book *The Immunologic Theory of Aging* published in 1969. He argues (like Burnet) that aging is due to an autoimmune process that results in the immune system attacking its own tissues. This may come about in Walford's view either because the immune system becomes less accurate in its ability to discriminate 'self'; or because the cells of the body start displaying 'faulty' self molecules. Walford's theory can perhaps be said to be more wide ranging than Burnet's since anything that impairs recognition between immune system and 'self', will in his view, lead to immunological decline. This process may therefore involve all the various cells of the immune system (Burnet's theory tends to emphasise T-cells) and have multiple causes (not only somatic mutations but other types of cellular damage as well). Perhaps the biggest difference between Walford's theory and that of Burnet, however, is that Walford does not appear to believe that the thymus is the critical pacemaker of immunological decline. Instead, Walford sees the decline as having multiple causes and occurring throughout the immune system. Consequently, Walford's view of aging can perhaps be best described as *pathogenic* (or 'disease-like') rather than programed.

THE GRAFT-VERSUS-HOST REACTION

Walford has also been instrumental in providing experimental evidence to support his theory. In particular, he has attempted to mimic what he believes is happening in aging by inducing what is known as a 'graft-versus-host' reaction in laboratory animals. As is well known, our immune system will normally reject the tissues taken from another individual following organ or skin transplantation. But in some cases the roles can be reversed so that the new grafted tissue actually attacks the host. This often happens, for example, in the case of bone marrow transplants. When bone marrow is harvested from a donor it is impossible to avoid including large numbers of T-cells along with the marrow cells; and consequently when the T-cells are placed into the recipient, they typically regard their new home as a gigantic transplant and go about attacking it (even today with advanced drug therapy bone marrow transplants have a high risk of mortality for this very reason). To mimic this effect, Walford has implanted white blood cells from adult mice into genetically similar (but not identical) newborn mice. In this type of study, the new-born mice are not old enough to recognise the transplanted cells as foreign, but the transplanted white cells are mature, and consequently they set up a 'graft-versus-host' reaction. Of course, this operation would normally lead to the quick death of the animals, but by using special strains of mice that are almost (but not quite) genetically identical, Walford has managed to set up a slow and delayed autoimmune reaction – and one that he predicted should resemble aging.

Indeed, the results of such studies have tended to confirm Walford's predictions with the transplanted mice showing many signs of accelerated aging including early loss and greying of hair, increased vascular and kidney disease, weight loss, amyloidosis and an increase in autoantibodies. Walford also reports that collagen ages much faster in the transplanted animals. Moreover, mice transplanted with foreign white blood cells tend to have shorter lives (although this effect may be partly due to an increase in lymphoma cancer). Overall, Walford's work appears to support the immunological theory of aging, and shows that autoimmunity can cause accelerated aging.

EVALUATING THE IMMUNOLOGICAL THEORY

The immunological theory has been the subject of a great deal of research but, unfortunately, it has produced a somewhat fragmented picture. One reason is that the basic idea of aging being caused by a long term, and relatively weak, autoimmune response is very difficult (if not impossible) to measure directly in any given individual. Because of this, one has to infer its existence experimentally rather than to see the process 'at work' directly.

Perhaps the most controversial issue surrounding the immunological theory of aging is the role of the thymus gland. In short: does the thymus gland acts as a biological clock regulating the decline and efficiency of the immune system (as Burnet maintains), or does it have little role to play in the aging process? In support of the former idea, there appears to be some evidence that the thymus does indeed have 'clock-like' properties. For example, Burnet has shown that when a thymus is removed from a young animal and implanted into an older animal, it grows as a young thymus apparently unaffected by the age of its new host. Moreover, according to Burnet, one can transplant a dozen, or more, thymuses from young mice under the skin of an adult and all will grow to normal size, unaffected by the fact that there is an abnormal amount of thymic tissue in the body. This suggests that the thymus may have its own intrinsic clock which would make it an ideal candidate for a biological time-keeping role.

However, if this theory is correct, then one would expect the removal of the thymus gland to have an important effect on immune function and aging. But the evidence on this point appears to contradict Burnet's theory since removal of the thymus only has life shortening effects when undertaken in new-born animals, and not in young or mature animals. It is difficult, therefore, to reconcile these findings with the view that the thymus is the so-called pacemaker of aging.

Despite this, there are reports of accelerated aging in animals that have a congenital failure of thymus development. For example, two genetic strains of mice (called 'Nude' and 'Snell-Begg') are known to be born without a proper thymus and both show abnormalities of aging. The Nude strain are as their name suggests hairless, but more importantly they show no T-cell function (i.e., they readily accept foreign tissue grafts), and they normally die within four months of birth. During this time they undergo a number of aging-type changes including rapid loss of body weight, thinning of skin, loss of cutaneous fat and reduced body metabolism. But this can be reversed by the simple transplantation of a new thymus gland – and following the operation, the improved health of these mice is quickly apparent. They begin to gain weight and their general vigour and health improves. In fact, some of these mice have lived and remained healthy for well over 550 days.

In contrast, Snell-Begg mice are often referred to as hypopituitary dwarf mice because they are born without a pituitary gland, and one consequence of this abnormality is a deficiency of growth hormone which results in them being extremely small without a properly formed thymus gland. Not only do these animals rarely live beyond 4–5 months, but they also show signs of accelerated aging including rapid greying of hair, atrophy, and wasting of body tissues. One way of arresting this decline is to inject these mice with mature lymphocytes (i.e. T-cells) derived from a genetically similar but normal mice, thereby helping to reinstate the function of their immune system. Remarkably, the injection of lymphocytes has been found to increase the life span of Snell-Begg mice to around 15 months, whilst abolishing their age-related pathology. In other words, the administration of lymphocytes slows down their aging decline, thus lending support to the view that the thymus, or rather, the immune system in general, may be involved in the process of aging.

It seems to follow from this work, that it should also be possible to rejuvenate old individuals by injecting them with immune cells taken from young individuals. To do this one needs to take an aged animal and destroy the source of its immune cells (i.e., its bone marrow): and then inject into the recipient a new and 'youthful' suspension of new bone marrow cells. Bone marrow, of course, contains the stem cells that give rise to new lymphocytes (and ultimately T-cells), so this type of transplantation should rejuvenate the 'old' immune system, and (if the immuniological theory is correct) increase the life span of the recipient. Indeed, this type of operation has been undertaken, and providing a new thymus is also transplanted into the old animal, it has been shown that the efficiency of the immune system in old animals can be restored to that found in young animals. However, despite this, it has also been found that old mice treated in this way do not live any longer than untreated animals, and neither do they show any differences in the speed of their aging providing evidence against the immunological theory.

THE MAJOR HISTOCOMPATIBILITY COMPLEX AND AGING

We have already seen that a crucial requirement of the immune system is the ability to distinguish *self* from *non-self* which it does by recognising a special set of 'self' proteins found on every cell of the body called MHC proteins. MHC stands for major histocompatibility complex, and this refers to a cluster of some 40–50 genes (that in humans are found on chromosome 6) which are the coded instructions for making the MHC proteins found on the surface of our cells. Because there are at least 50 slight variations (alleles) of each gene, the possible variation in MHC proteins is so great that it is virtually impossible for any two people, except identical twins, to have matching sets of MHC proteins (in fact, the chances of finding somebody with identical MHC proteins is about 1 in 20 million). This great variation helps explain why the immune system rejects foreign tissue following transplantation between individuals.

One of the main functions of the MHC proteins is to present antigens to the lymphocytes and to help direct the control of the immune response. For this reason, Roy Walford has reasoned that if the immune system is involved in aging, then it is probable that the MHC genes are somehow involved in this process. Working with George Smith at the University of California, Los Angeles, Walford tested this theory by examining the life spans of different strains of cogenic mice. These are mice that have been bred by brother–sister matings through multiple generations until all the individuals of any one strain are like identical twins, except in this particular case the mice only differed in terms of their MHC genes. In other words, given these strains of mice, any difference in life span could only be explained in terms of MHC gene differences. Smith and Walford tested 14 different strains of mice and found that their average life spans ranged from 85 to 141 weeks. In short, the genetic inheritance of MHC genes was shown to have an important effect on determining longevity.

Interestingly, there is also some human evidence showing that differences in MHC genes can affect longevity. For example, Hajime Takata and his colleagues working in Tokyo, have examined the MHC genes of 82 verified centenarians who lived on the Okinowa Islands of Southern Japan. These islands are well known for their long lived inhabitants and in 1984 about 5 per 100,000 of the population were over a hundred years old. These researchers found that one MHC gene (called DRw9) was especially rare in the centenarians and was found in only 8% of individuals (compared to 31% of normal Japanese). In fact, this particular gene is often found in oriental people suffering from a range of autoimmune disease including systematic lupus erythemetosus, diabetes and myasthenia gravis. Thus, the DRw9 gene appears to be a considerable risk factor for good health and its absence a favourable factor for longevity. In addition, another gene (called DRI) was not found at all in the 159 mainland Japanese but was present in 6% of centenarians. Thus, there may be a genetic factor in the longevity of the Okinowan islanders (although there are also important lifestyle factors and these will be discussed in the next chapter).

Also supporting the immunological theory is the fact that there are certain genes of the MHC complex that are known to increase the risk of certain diseases of aging including rheumatoid arthritis and diabetes. In addition, the MHC also appears to have a wider range of functions than just being involved in immune regulation. For example, the gene regulating superoxide dismutase activity, which helps protect cells against the effects of free radicals (see pages 49 and 115–117), has been found to be genetically linked with the MHC. Moreover, there is some evidence that MHC genes are also involved in DNA repair processes. Thus, the MHC has a broad range of functions and Walford has speculated that in evolutionary terms it may actually represent a very ancient gene cluster whose original function was to protect the organism from free radical damage. However, as evolution has progressed, it has had functions added to it which are associated with aging, including regulation of immune function. Whether this is true, or not, Walford's own research clearly suggests that the genes of the MHC have a very important influence on both the immune system and life span.

AGING AND AUTOIMMUNITY

If aging is linked to autoimmunity then we might expect the incidence of autoimmune diseases to increase with age. However, with a few exceptions, this relationship does not appear to hold true. The fact that humans are susceptible to a wide range of autoimmune diseases is beyond doubt. These disorders can affect practically every organ system of the body including the brain and spinal cord in multiple sclerosis; the lining of the joints in rheumatoid arthritis; the connections between nerve and muscle in myasthenia gravis; the thyroid gland in Graves' disease; and the skin in pemphigus vulgaris. Other types of autoimmune disorder such as lupus ertthymatosus can affect tissue throughout the body causing widespread damage. Autoimmune diseases are, in fact, fairly common. It has been estimated that approximately 5% of the adult population suffer from an autoimmune disease, with about two-thirds of all victims being women. The fact that females are much more prone to immune diseases, yet typically are longer-lived than males, doesn't appear to equate with the immunological theory of aging.

Furthermore, although many autoimmune disorders begin relatively early in life (between 20 and 40 years) they generally do not increase in a parallel fashion with aging. A good example is multiple sclerosis. Onset of this disease is rare before puberty, it reaches a peak between 30 and 40 years, and then declines sharply so that its onset is rare after the age of 60 years. In fact, each autoimmune disease appears to have its own characteristic pattern. For example, systemic lupus erythematosus often begins in early adulthood, rheumatoid arthritis reaches its peak of prevalence in the 55 to 65 age group, and thyroiditus is most common in people aged 60 to 70. Thus, it is difficult to view these disorders as diseases of aging.

Are there any immune disorders in prevalence with aging that continue to increase with aging? According to Roy Walford there are several including inflammation of the arteries, certain kinds of anaemia, maturity-onset diabetes and amyloidosis. The latter is particularly interesting because it results in the deposition of amyloid, a hard starch-like substance that accumulates in the small blood vessels (including those of the brain) as well as in various organs of the body. Amyloidosis also tends to be one of the most characteristic features of old age. However, overall, the list of autoimmune diseases that Walford cites is hardly impressive, and moreover there is considerable doubt whether amyloidosis is really an immune phenomenon. In short, the evidence from autoimmune diseases does not appear to provide convincing support for the immunological theory.

Another way of examining the issue of autoimmunity and aging is to measure the number of antibodies that have become sensitised to the body's own tissues (autoantibodies). A large number of autoantibodies can now be detected, and if the immunological theory is correct then we would expect that their number to increase with aging. Indeed, a number of studies have shown this to be the case. One of the more impressive studies examining this issue was undertaken by Dr B. Hooper and his colleagues from the University of Melbourne, who in 1969 measured the levels of autoantibodies in over 90% of the adult population of the Western Australian town of Busselton (some 3,500 persons). The results showed that the level of most types of autoantibodies in the blood tended to increase significantly with aging. For example, approximately 15% of 20 years old subjects showed evidence of autoantibodies, and this figure rose to around 40% in those aged over 70 years. Despite this, the Busselton study showed that the *majority* of people did not show evidence of autoantibodies even in old age, and therefore they could not be used as an explanation of aging.

The relevance of autoantibodies to the understanding of aging, however, is still somewhat controversial. Some experts believe that they are the cause of several age-related illnesses such as vascular disease whereas others believe that autoantibodies form a normal part of the immune response, or are a normal response of the organism to rid itself of damaged tissue by inducing the metabolic breakdown of the cells. Put simply, their significance is not fully understood and their relevance to aging unproven. In fact, such a statement could also probably be said about the immunological theory in general. Thus, whilst the great merit of the autoimmune theory of aging is that it provides a general explanation for the many divergent features of aging; the complexity of the theory unfortunately seems to have rendered it, for the time-being at least, unproven one way or the other.

SUMMARY

- Important structures of the immune system include the bone marrow, thymus, spleen and lymph nodes. The main cells of the immune system are the various types of white blood cell (including lymphocytes) that are to be found in the blood and lymph.
- The first line of defence for the body is natural immunity. This includes protective barriers (such as skin and mucus membranes), macrophages (mainly found in the lymph nodes) and neutrophils (found in the blood).
- The immune system is able to 'acquire' an immune response by the production of antibodies (from B-cells) which can also be used as an immunological memory should the antigen ever be re-encountered.
- Another type of acquired immunity involves the production of T-cells that are designed to recognise specific antigens. T-cells are particularly important as they are able to recognise (and destroy) infected cells of the body.
- The most striking age-related change that takes place in the immune system is the thymus gland. It reaches its full size in late childhood and then declines so that by the age of 50 years it only retains some 5–10% of its original mass.
- There appears to be a significant decline in the function of T-cells with aging (cell mediated immunity) although there does not seem to be a marked decline in the function of B-cells (antibody immunity).
- The main idea behind the immunological theory is that the immune system becomes less able to distinguish *self* from *non-self* and therefore gradually starts to attack its own body, resulting in the changes we recognise as aging.
- Evidence for the immunological theory has come from studies looking at the graft-versus-host reaction in which foreign marrow cells are implanted into new-born mice. These animals show many signs of accelerated aging.
- Further evidence for the immunological theory has come from cogenic mice that are genetically identical except for alleles in their

MHC genes. Smith and Walford examined 14 different strains of such mice and found that their life spans ranged from 85 to 141 weeks.

• There is also evidence at odds with the immunological theory. For example, removal of the thymus gland only shortens life span if performed in very young animals; and moreover it is not clear if autoimmune disease really does increase with aging.

CALORIE RESTRICTION AND LIFE EXTENSION

To lengthen thy life, lessen thy meals

–Benjamin Franklin

If dietary restriction has the same effects in humans as it has in rodents, then human life span can be extended by at least 30 percent – which would give us an extra 30 to 35 years.

–Edward Masoro

Just as survival is one of the most basic instincts of the animal kingdom, the wish to extend one's life expectancy is one of the strongest desires of human beings. We all like to think we can live to a grand old age, and maybe a bit more besides, although realistically we are also likely to be only too aware of the limits of our existence. Sadly, whatever we may do to maximise our chances of a long life, there is little evidence that any form of life extension regime currently being practised by human beings will be able to push our life span beyond its maximum of about 110–120 years. Despite this, it might come as a surprise to find out that scientists have known for over 60 years that life extension by up to 50% is possible in laboratory animals, and that the procedure is simple, robust and reliable. The method is calorie restriction, or more specifically, reduced feeding without malnutrition, and no other form of life extension comes remotely close to it in terms of its effectiveness. Indeed, because it is such a reliable means of extending longevity, most researchers agree that understanding how calorie restriction works will not only provide invaluable clues about the causes of aging, but should also enable us to find ways of slowing down its decline. Consequently, calorie restriction is one of the most active research areas in biological gerontology today. However, even without scientific explanations, the benefits of the procedure are there for all to see, and if we want to maximise our chances of a long life, we ignore the lessons of dietary restriction at our peril.

As we saw in the first chapter, the Gompertz decline in life expectancy (i.e., the doubling of mortality ever 8 years or so) only comes into play once we reach adulthood. The reason for this is that during childhood and adolescence we 'age' in terms of years, but we do not 'age' in terms of biological decline. It obviously makes no biological sense to 'decline' as we grow, and thus aging (senescence) is something that only gets put into motion once our growth (or sexual maturation) has finished. This raises the interesting implication, however, that if we could find a way of continuing to grow throughout life, even if very slowly, then we might be able to stop aging. In fact, there does appear to be some evidence for such an idea. For example, many fish (such as flatfish and sturgeon) and a number of big reptiles (including the Galapagos tortoise) experience 'indeterminate' growth whereby they continue growing for as long as they live. And, indeed, such animals often grow to enormous sizes and live to great ages. This does not mean that these animals are immortal, because like us they are also prone to fatal accidents and disease. But, nonetheless, they do not show any marked biological decline with aging; and their death seems to be more a matter of chance, rather than something that is bound to occur in a Gompertzian fashion with increasing age.

It was with this in mind that Clive McCay working at Cornell University in the 1930s began his classic experiments looking at the effects of calorie restriction in rats. McCay reasoned that if there was a link between growth and longevity, then by slowing down and prolonging the period of growth in laboratory animals, it should be possible to increase their life spans. To do this, McCay set out to feed his animals as little as possible in order to retard growth, but at the same time provide enough nutrition to maintain their health. McCay began his experiment with 106 white rats which were divided into 3 groups just after weaning. The first group of animals were allowed unlimited access (*ad libitum*) to laboratory food which enabled them to grow normally. The second group were placed on a special diet developed by McCay, that was balanced with all the necessary vitamins and minerals – except the calorie content of the food was dramatically reduced to a semi-starvation level (about one-third of normal). The third group was allowed two weeks of unlimited access to food following weaning enabling a certain degree of normal growth to take place, before they were then also transferred to the restricted diet.

To achieve the lowest rate of growth possible, McCay used what was called the "stair-step" method of feeding. This technique involved keeping the animal at a constant weight for between one to four months, and then allowing a small weight gain of ten grams by increasing the animal's allowance of food. This approach to weight control turned out to be very successful since McCay found that his rats remained small and most continued to grow throughout their life span, unlike the rats given unlimited access to food. In fact, after 18 months when normal rats had been fully grown for more than a year, the starved animals were only about one quarter size, and none ever reached the full grown size of their fully fed companions. However, the most striking finding of all was the marked difference in life span between the groups. Remarkably, the results showed that calorie restricted rats lived nearly 50% longer than the control animals. For example, the normal animals lived on average for 483 days compared to 820 days for the food restricted rats. In terms of maximum life expectancy, the results showed that the longest-lived *ad libitum* fed rat managed to survive 969 days compared to 1465 days for the longest-lived restricted animal. Thus, McCay had not only increased the average life span of his animals, but more importantly he had also increased their *maximum* life span.

Even more striking was the healthy demeanour of the calorie restricted animals. These animals at the age of 1,000 days (by which time all the normal animals had died) had the appearance of much younger rats. They had glossy sleek coats, were active and mobile, and had superior levels of intelligence. Indeed, as McCay himself noted, despite their extreme old age, the food deprived rats had the appearance of young healthy animals. In fact, some researchers believe that McCay may have understated the effect – for example, Robert Prehoda in his book *Extended Youth* argues that McCay's 1,000 day old retarded rats were equivalent to a 90 year old human having the physical appearance of a teenager! In short, such animals appear to be in a state of persistent youth and without any sign of frailty. McCay's findings have now been replicated hundreds of times by a number of investigators, and in a number of different species including animals as diverse as protozoa, water fleas, spiders and fish. Research is currently underway with primates, and preliminary findings indicate that dietary restriction is having beneficial effects in these animals as well – suggesting that it may also be applicable to humans.

THE EFFECTS OF CALORIE RESTRICTION ON AGING

Research has shown that the main reason why calorie restriction extends life span is because it slows down the rate of aging. As we have seen throughout this book, almost every biological system in the body shows a decline in function with aging, but in nearly every instance it has been found that the extent of the decline is less in animals that have been food restricted. Over the last 50 years or so, biologists have examined the effects of dietary restriction on over 300 biological parameters that are known to decline with aging, and it has been found that approximately 90% of these are functionally more 'youthful' in food restricted animals. In other words, animals placed on restricted diets are indeed physiologically younger than their actual age suggests.

One of the most striking changes that takes place as a result of calorie restriction is the more youthful appearance of the body. Although food deprived animals tend to be smaller than normal, they nevertheless appear to be fitter, leaner, and more muscular. This has also been verified in research that has looked at the body composition of these animals. For example, in one study that looked at fully fed rats it was found that the protein content of these animals declined by 44% between the age of 12 and 28 months of age, whereas the fat content increased by some 43%. Thus, with rodent aging, muscle mass declined at the expense of fat (this change also occurs with humans). However, the body composition of the food restricted animals did not follow the same trend. Although these rats had smaller amounts of protein and fat, the relative proportions of these two components did not change with age. In fact, the relative proportions of fat and protein was the same at the very old age of 28 months as it was at 12 months.

As we saw in chapter 5, one of the best ways of estimating the biological age of an individual is to measure the strength of their collagen fibres. As individuals age, their collagen (which forms an important component of connective tissue) becomes more rigid because it forms increasing crosslinks that tie the fibres together. Consequently, collagen fibres become stronger but less flexible with aging. Biologists can examine the tensile properties of collagen by removing a sample of it from the body and measuring how much force is needed to break the fibres: or alternatively, measure how long it takes for collagen fibres to dissolve in salt baths. Either way, research shows that there is a marked increase in collagen rigidity in normal animals that correlates well with their age. However, the strength of collagen is significantly reduced in food deprived animals showing that they are indeed biochemically younger than normally fed animals.

But the benefits of food restriction go far beyond the slowing down of collagen aging. For example, animals placed on restricted diets show less degeneration of heart tissue, and have an almost non existent incidence of atherosclerosis probably due to reduced cholesterol levels in their blood. They have improved kidney and liver function, and demonstrate improved immune functioning. In addition, there is little decline in muscular strength or neural degeneration even at advanced ages. Similarly, DNA repair, glucose tolerance, protein synthesis and hormone action all show less decline in food restricted animals. Despite this, not all age-related changes are modified by food restriction since it does not appear to slow the deposition of lipofuscin in cells, change blood pressure or modify the activity of certain enzymes. Thus, it may not be totally accurate to say that an aged food restricted animal is biologically identical to a younger animal, although it certainly comes close.

One of the most interesting consequences of food restriction is its effect on reproductive development. For example, one study showed that while normal female rats began their oestrous cycles at around 2 months of age, this event in food restricted animals did not occur until some 5 months later. However, one benefit of this delay was that food restricted animals were able to successfully reproduce at much older ages. For example, B.J. Merry and Anne Holehan working at the University of Hull, have shown that while most normal rats have stopped their oestrous cycles by 18 months of age, nearly two-thirds of dietary restricted animals were still having oestrous cycles 6 months later. And in most cases these animals were fertile and able to bear litters (in fact, 25% of animals were still able to breed past the extreme old age of 26 months). The benefits in reproductive function also extended to litter size. Whilst fully-fed females showed a decline in litter size with increasing age, this did not occur with the underfed mothers who continued to produce large litters throughout their reproductive life span.

Surprisingly, the benefits of dietary restriction on male reproductive function is not as marked. Food restriction delays the developmental peak of testosterone by about 20 days, and there are lower levels of this hormone in adulthood. The delay in maturation in dietary restricted males is also reflected in their ability to sire litters. For example, it was found that between 63–84 days of age, only 30% of restricted males sired one or more litters compared to 90% in the fully fed group. However, by the time both sets of animals had reached 100 days of age there was no difference in fertility between the two groups; and following this point, the age-related decline in fertility appeared to be similar in both groups. Thus, compared to females, food restriction does not appear to have such a significant effect on slowing down reproductive decline in the male.

Not only does food restriction slow down the rate of aging, but it also delays the onset of age-related diseases and alters their incidence. This was first shown by McCay and has been confirmed countless times since. In McCay's case the main cause of death in both his fully-fed and restricted groups was pneumonia. McCay showed that the likelihood of this disease increased with age in both groups, but it occurred much later in the underfed animals. Thus, food restriction delayed, rather than prevented, the occurrence of pneumonia, showing if nothing else that the restricted rats had stronger immune systems than fully-fed animals. However, this disease must be regarded as a premature death since it can only occur if the pneumonia virus is able to establish itself in the rat colony and today, improved husbandry conditions have helped overcome this problem allowing animals to die from more 'natural' causes.

Laboratory rodents typically die from a number of diseases, but one of the most common forms of death in most strains (although not all) is kidney damage, or what is sometimes called glomerular sclerosis. Glomeruli are tangles of blood vessels in the kidney that are involved in filtering fluid to form urine. With aging these tend to show marked degeneration and inflammation, often leading to increased blood pressure and changes in the mineral and acid balance of the blood. In one study undertaken by Morris Ross at the Institute of Cancer Research, in Philadelphia, it was found that among well fed rats, 40% developed in varying degrees of severity, some form of kidney damage. However, for animals placed on restricted diets, this condition was rare and in the few instances when it was found, the disease occurred at advanced ages and was less severe. These findings may have relevance to humans. Although humans do not normally die from glomerular sclerosis they nevertheless often show a decline in glomerular function with aging which may have important health implications – such as increased blood pressure.

However, perhaps most impressive of all is the effect of food restriction on delaying the development of tumours. Rodents are prone to a wide range on different types of cancer, including lung, mammary gland, ovary, pituitary gland and lymphomas; with certain strains of rats and mice being more susceptible to particular types of cancer than others. However, food restriction significantly reduces tumour incidence, and the more severe the underfeeding the greater the protection. In fact, even quite moderate levels of food restriction can have beneficial results. For example, Morris Ross has found that in the case of mammary and liver tumours, restricting rodent intake of food by around 35% can reduce the incidence of these tumours by nearly 100%. Some studies have obtained striking results with diets that have only used 20% food restriction. For example, a study published by Mary Tucker in 1979 found that 80% of fully-fed mice eventually developed tumours at some point in their life (on average at 20 months); whereas only 50% of food restricted animals developed tumours, and these occurred much later (at around 30 months). Thus, there appears to be little doubt that food restriction has a major impact on slowing down the onset and development of tumour formation.

CALORIE RESTRICTION AND GROWTH

The important question therefore is: how does calorie restriction work? An implicit assumption made by McCay was that it worked because the starvation of food delayed the growth of the animal. That is, because the food deprived animal grew slowly throughout its life, the continuation of growth protected against the aging process. This, of course, leads to the prediction that dietary restriction should be most effective if implemented early in life before the animal reaches full size. Thus, if McCay was right then placing animals on restricted diets once they had reached their full size should have no benefit. However, a number of studies have shown this not to be the case. In fact, some studies that have allowed rats to grow normally and reach maturation before imposing calorie restriction, have shown that such animals live almost as long as those that have been restricted throughout their life span. In short, evidence suggests that it is not the slowing of growth that is the important factor in dietary restriction, but rather some consequence of the diet itself.

What is the best time in the life span to implement dietary restriction? This question has been examined in studies that have switched animals at different ages from restricted diets to *ad libitum*, and vice versa. Unfortunately, most studies have used varying degrees of dietary restriction which have made the results difficult to compare. Nevertheless, the general consensus is that food restriction started early (after weaning) and maintained throughout life has the most beneficial effect of all. However, it is also probably the case that restriction for any part of the life span (unless occurring in old age), has positive effects – providing the period of the deprivation is long enough. For example, Anne Holehan and B.J. Merry returned rats to normal feeding after 7, 39, 159, 236 and 344 days of restricted diets, and showed that approximately one year of underfeeding was necessary to obtain a significant (10%) extension of maximum life span. Since the rats they used (Sprague-Dawley) typically have a maximum life span of 900 days, extrapolating this data to humans would seem to suggest that we would have to undernourish ourselves for about 30 years to obtain any real benefit in life extension.

However, there appears to be an alternative and less harsh way of achieving the same aim. In 1946, Anton Carlson and Frederick Hoelzel working at the university of Chicago, examined the effects of intermittent fasting on the life span of rats. They fasted groups of rats 1 day out of every 2, 3 or 4, and discovered that fasting 1 day in 3 increased the average life span of male rats by 20% and females by 15%. Interestingly, the animals that had been fasted 1 day in 4 also showed some degree of life extension despite having *ad libitum* access to food in the non-fasting days. Furthermore, all groups of animals demonstrated normal growth and weight for their age, again showing that the effects of calorie restriction are not due to slower growth. This effect has also been demonstrated more recently by Charles Goodrick at the National institute of Aging. In fact, all this evidence has so impressed Roy Walford (see previous chapter) that he personally follows, and recommends, total abstinence from food for two successive days each week, whilst consuming a healthy diet for the other 5 days. At present he is in his mid 70s and his dietary regime appears to be working well.

WHAT DIETARY COMPONENTS ARE IMPORTANT?

It goes without saying that all animals require a balanced diet containing carbohydrates, protein and vitamins for survival. Carbohydrates in the form of sugars and starches are the main energy source of the body; and these are typically broken down into glucose, which can either be stored in the liver and muscles for immediate release, or transformed into deposits of fat. In contrast, proteins are needed for growth and tissue repair, the importance of which can be readily shown by the fact that children, in relative terms, need two or three times more protein than adults. It is sometimes said that carbohydrates provide the energy, and proteins provide the substance of the body, although strictly speaking this is not true as protein can also be broken down (if needed) by the body to provide calories and energy. Fat provides a third component of diet acting both as a source of energy and a structural material. Thus, carbohydrates, proteins and fats provide the calories in our diet.

We also require a number of vitamins and minerals to enable metabolism to take place although these chemicals cannot be broken down for energy and therefore have no caloric value.

An important question regarding calorie restriction concerns the relative importance of the dietary components in the restricted diet. In other words, is the increase in longevity due simply to the restriction of calories – or is it due to some component of the diet such as protein or carbohydrate? Initially, McCay believed that it was the lack of protein in his diet that was the most important factor since this was the vital component needed by the body to grow. However, McCay never adequately tested his theory, and it was left to others to examine this issue more thoroughly. One of the most extensive studies to examine this question was undertaken by Morris Ross in 1959. Using large groups of animals (over 1,600 in total) Ross looked at the effects of calorie restriction on the longevity of laboratory rats, but manipulated the relative proportions of protein and carbohydrates in their diet. Of particular interest was the comparison between animals placed on the high protein-high carbohydrate diet and the low protein-high carbohydrate diets. Both these diets were designed to contain exactly the same amount of calories, but to vary in terms of their protein content. In contrast to what McCay might have predicted, Ross found that the animals placed on the high protein diet actually lived longer than the animals placed on the low protein diet. Moreover, these animals also grew faster and reached a greater maximum weight than the low protein animals (440 g compared to 394 g).

Yet again, this work shows that the beneficial effects of calorie restriction is not linked to growth, and if anything, a high protein diet (that encourages growth) is superior to a low protein diet. However, others have not always been able to confirm these findings, and some have even reported that high carbohydrate diets are better than high protein diets. More recent studies, however, have tended to show that the important factor lies with the amount of calories, and not any particular component of the diet. For example, a study undertaken by Teresa Davis and her colleagues in 1983, gave calorie restricted rats a diet that contained the same amount of protein as fully fed animals, but with a third fewer calories. The results showed that the deprived animals lived longer despite having consumed the same amount of protein as fully-fed rats. Research manipulating other dietary components in the same way have tended to provide similar findings, and consequently most researchers now accept that the most important factor of all, in diet restriction, is the reduction of calories (no matter what source they come from). The implication, therefore, is that our longevity is determined by the total amount of food eaten, measured in terms of calories, and not by any particular constituent of our diet.

HOW DOES CALORIE RESTRICTION WORK?

Not surprisingly, biologists have made considerable efforts to understand how dietary restriction is able to slow down aging and delay the onset of disease. Indeed, understanding this process is likely to unlock many of the secrets of aging and to offer hope in its intervention (for example, it may be possible to develop a drug that will mimic the effects of food restriction). However, as yet, nobody knows for sure how food restriction works although as we shall see there are no shortage of theories.

'COMMON SENSE' THEORIES

The original theory offered by McCay was that food restriction worked because it slowed down the rate of growth and development during the early period of life. This theory makes a great deal of intuitive sense because it is difficult to imagine an animal aging whilst it is still growing. Consequently, by stretching out the maturation period, McCay believed that he was extending the life span of the animal. However, this theory is now known to be incorrect for the simple reason that it predicts that food restriction will have the greatest effect on extending longevity when started before the animal is fully grown. However, as already mentioned, dietary restriction is effective even when it is initiated in adulthood. For example, Byung Yu working at the University of Texas has shown that restricting rats at 6 months of age (at which point they are fully grown) is almost as effective at increasing longevity as that starting at 6 weeks. Moreover, the pattern and onset of diseases in both groups were very similar. Yu has also shown that food restriction limited solely to the period of growth and development, is much less effective at increasing longevity than that started in adulthood but extending throughout the rest of the life span. Thus, it can safely be said that food restriction does not work because it slows down growth.

Another early theory was that food restriction works because it reduces the amount of body fat. This theory was originally proposed on the basis that excess body fat and obesity was known to be linked to premature mortality and disease in both man and animals; therefore it made sense to suppose that the relationship worked the other way, and that less fat helped to increase longevity. However, the common sense view has again been shown to be incorrect. Contrary to expectation, it has been demonstrated that it tends to be the *heaviest* animals in any given food deprived group that live the longest and not the lightest. Thus, in calorie restricted animals increased fat stores are actually advantageous for survival indicating that reduced fat levels cannot explain why food restriction works.

Another popular explanation is the rate of living theory (see chapter 7). As we have seen above, the important factor in diet restriction appears to be the total number of calories consumed rather than the composition of the diet itself. This finding would therefore seem to lend support to the theory first proposed by Rubner in 1908, and later extended by Raymond Pearl, that it is the rate at which an organism uses up its energy that is the critical factor in determining the length of life. In theory, therefore, an animal could use up its quota of energy quickly and shorten its life span – or utilise its energy more slowly and live longer. Linking this idea to calorie restriction, it would seem plausible to assume that a starved animal is being made to use up its energy quota more slowly – and therefore reduce its rate of living.

The speed at which an animal utilises energy can be examined by measuring the metabolic rate, or more precisely its rate of oxygen consumption (see chapter 7). However, comparing the metabolism of fully-fed and deprived animals is not a straightforward task as they contain different proportions of protein and fat which have different metabolic rates. Nevertheless, early research taking into account these differences, indicated that the metabolic rate of food deprived animals was indeed lower than normal. Further research also showed that restricted animals tended to have a lower body temperature – providing further support for a decreased rate of metabolism. And, in 1977, George Sacher provided yet more evidence. He re-examined the results of an earlier study by Morris Ross that looked at the effects of 5 different dietary regimes which ranged in calorie intake from 18 to 75 kcal per day, and produced survival times ranging from 780 to 990 days. When Sacher calculated the total calorie intake per gram of body weight during the life span of all the animals, he found that calorie intake was practically the same for all 5 groups – with each animal using up approximately 102 kcal for each gram of its body weight across its life span. In other words, the lifetime caloric consumption was nearly constant for all animals which strongly supported the rate of living theory.

Despite this, recent work has cast serious doubt on the rate of living theory as an explanation for calorie restriction. For example, Roger McCarter and his colleagues at the University of Texas, examined metabolic rate (by measuring oxygen consumption) for 23.75 hours each day over a long term period (4.5 months) for groups of rats that were either fully-fed or calorie restricted. Surprisingly, McCarter found no difference between the two groups in their oxygen consumption, despite marked differences in their weight. Further work by McCarter showed that although there was a short term decrease in metabolism, it was transient and occurred only over the first 6 weeks of the deprivation period. Beyond this point the metabolic rate of the deprived animals reverted back to normal, and was the same as the fully-fed animals.

Even more convincing evidence against the rate of living theory was provided Edward Masoro, also at the University of Texas, who placed one group of animals on an unlimited diet, and then placed a second group on the same diet, except that they received only 60% of the amount eaten by the first group. This regime clearly had a beneficial effect on longevity since the restricted animals lived on average for 986 days compared to 701 days for the *ad libitum* group. Moreover, the restricted animals, as might be expected, were significantly lighter than their fully fed counterparts. However, when Masoro compared the number of calories consumed over the lifetimes of the two groups, and adjusted the values to account for the differences in the body weights, he found that the restricted rats actually consumed more calories per unit of body weight than the fully fed animals! In other words, this was an opposite finding to Sacher's and showed that the rate of living theory did not fit the data after all.

Do these findings rule out a metabolism-type explanation? Not necessarily. It might be that food restriction, rather than reducing metabolism, causes a change in the way the body utilises energy. For example, food restriction results in a shift away from fat synthesis towards glucose synthesis, and it is now known that restricted animals have a normal metabolic rate prior to feeding, but a higher than normal metabolic rate after feeding. In addition, a wide range of enzymes are altered by dietary restriction including those in the liver concerned with drug metabolism and elimination. Thus, it appears that animals on dietary restriction have an altered metabolism which does not allow meaningful or simple comparisons with normal animals – or indeed with a simple rate of living theory.

THE FREE RADICAL THEORY

One of the most popular current explanations of how calorie restriction works is its effect on reducing free radical damage. As we saw in chapter 5, free radicals are essentially atoms that contain an unpaired outer electron and, in chemical terms, this makes them very reactive with other biological molecules. Most free radicals are derived from oxygen (it has been estimated that between 1–3% of all the oxygen we breath is ultimately converted into free radicals), and the main site in the cell where they are formed are the mitochondria (see pages 161–165). All parts of the cell are susceptible to free radical damage, although most vulnerable are the mitochondria themselves, and this in turn, may interfere with the cell's ability to produce energy. And as the cells become less efficient, so do the tissues and organs of which they are part. But our cells are not defenceless against these harmful chemicals, and they have evolved a number of enzymes (including superoxide dismutase, catalase, and glutathione) that help protect against the constant barrage of free radical attack. Nevertheless, their protection is not complete and free radicals are considered by many to be a major cause of aging.

Several lines of research indicate that dietary restriction acts to reduce the amount of free radical damage that is taking place in the cell. For example, it has been shown that cellular levels of free radicals are significantly lower in animals subjected to long term calorie restriction, and this is associated with reduced amounts of damage to mitochondria and DNA – particularly in brain, heart and kidney cells. Perhaps even more important is the fact that dietary restriction also reduces the age-related decline in free radical enzymes. For example, levels of catalase and glutathione both normally decline with age, but these are maintained, or even increased, in animals undergoing food restriction. Thus not only are fewer free radicals being produced by dietary restriction – but perhaps even more importantly – the cell has an improved defence mechanism to cope with them.

HORMONAL THEORIES

As we have seen in previous chapters, the endocrine system is able to influence almost every cell and tissue in the body, and because of its pervasive role, it is perhaps not surprising that many investigators see it as having an important role over the body's decline. If this is true then we might expect to find that food restriction increases longevity by slowing down the aging of the endocrine system. One site where this effect could be expected to occur is the pituitary gland which controls the release of most hormones throughout the body. Indeed, as discussed in chapter 8, one way of increasing the life span of laboratory animals is to remove the pituitary gland, which reduces the levels of many hormones in the body. Thus, if food restriction exerts its anti-aging effect by acting on the endocrine system, we might expect that one site for its action would be the pituitary gland. Indeed, support for this idea comes from research by Arthur Everitt and his colleagues in Australia, who showed that there are many similarities between pituitary-lesioned animals and those placed on food restriction. For example, both types of animal show reduced aging of collagen fibres, a similar reduction in the number of age-related diseases, and an increase in life expectancy. In short, the effects of food restriction appear to mimic the effects of pituitary gland removal.

It might be predicted from this work that food restriction should also significantly lower the level of circulating hormones in the body, and in general (with one or two exceptions) this appears to be the case. For example, a number of studies have found reduced levels of thyroid hormones in food restricted animals (particularly triiodothyronine). As regards growth hormone, it appears that short term food restriction may suppress its release, although there is some doubt whether this decrease is maintained if under-feeding is continued over the long term (i.e., over 4–5 months). The situation regarding the sex hormones appears to be quite complex. In males, food restriction was found by B.J. Merry and Anne Holehan to increase levels of luteinizing hormone, but to marginally decrease levels of testosterone; whereas in females, both the oestrous peak of luteinizing hormone and oestrous cycles of estradiol was reduced. In general, therefore, food restriction appears to depress hormone release, although there is one notable exception: that is, the adrenal glucocorticoids, particularly corticosterone, which is found to be elevated in severely food deprived animals – possibly due to the stressful nature of the dietary manipulation.

How then does dietary restriction act to depress hormone release? One possibility may lie not with the pituitary gland – but with the brain structure that has direct control over the pituitary's release of hormones, namely the hypothalamus – which in turn is under the control of nerve cells that derive from other areas of the brain. Nerve cells send information to each other by releasing chemicals known as neurotransmitters, and one class of messenger found in the hypothalamus are the catecholamines which exert control over its release of hormones. It appears that with aging, levels of these catecholamines decline (which in turn would be expected to decrease the release of hypothalamic hormones and diminish their control over the pituitary gland). Moreover, Joseph Meites working at Michigan State University, has found that food restriction not only reduces the levels of catecholamines in the hypothalamus, but also slows down their age-related decline. This effect of food restriction is perhaps not surprising since catecholamines are made from the dietary amino acid L-tyrosine which would be reduced in underfed animals. The implication that follows from this, is that an important cause of aging may actually lie in the brain and, in particular, the hypothalamus. In fact, as we saw in chapter 9, a similar conclusion was reached by Caleb Finch who demonstrated that hypothalamic oestrogen exposure may be an important cause of reproductive decline in the female rat. Since the brain is the ultimate control system for practically every function of the body, then clearly a decline in its function could be an important factor in aging – as well as being a site where food restriction may exert its beneficial effects.

The hormonal theory of calorie restriction, however, is far from proven. One problem is that most investigators have shown food restriction to be a much more effective means of increasing maximum life span than pituitary gland removal. This is despite the fact that hormone levels (although depressed) are much higher following food restriction than with pituitary gland lesioning. The problem is also confounded by the fact that pituitary-lesioned animals typically eat less (normally about 50% less than as normal animals). Consequently it is possible that pituitary-removal has its beneficial effect on longevity because it also results in calorie restriction. Thus, a great deal of uncertainty still surrounds the hormonal theory of calorie restriction.

THE IMMUNOLOGICAL THEORY

Another possible explanation of why food restriction works is the immunological theory. The immune system continuously defends us against a large range of invading micro-organisms, as well as helping to protect us from our own cells that have become damaged or abnormal. As we have seen in the previous chapter, there is evidence to show that the function of the immunological system declines with aging – both in its ability to combat new antigens, and in being able to distinguish self from non-self. Thus, we might expect food restriction to slow down the age-related decline of the immune system and, in fact, this appears to be the case. The most obvious place where this occurs is the thymus gland. For example, Richard Weindruch has shown that at 6 weeks of age, the thymus gland taken from food deprived mice is only one-seventh of the size of that taken from fully-fed controls. However, at this age (the mouse equivalent to adolescence) the thymus in fully-fed mice begins to degenerate, whereas in the food deprived animals it continues to grow. Although the thymus will eventually start to shrink in the underfed mice, it is still significantly larger than that of normal mice by 6 months of age. In other words, the thymus develops more slowly in the deprived mice, but then remains larger for much longer in the adult before beginning to degenerate. The significance of this finding is still far from clear, although if the thymus acts a pacemaker to time the decline of the immune system (as some theorists believe), then this effect of food restriction may be a very important one in understanding how it slows down aging.

In support of this view, it has shown that the age-related decline in the efficiency of the immune system in underfed animals, appears to parallel the decline of the thymus gland. For example, the development of antibody responses and activation of T-cells in calorie restricted animals are slow to mature in the young, but then they show enhanced immunological responding for the rest of life (compared to normal). Further evidence for the immunological theory of dietary restriction has also come from a strain of mice known as New Zealand Black (NZB) which have a shortened life span (generally less than a year) because they suffer from autoimmune disease that results in an acute wasting of the muscles and internal organs. Remarkably, dietary restriction doubles the life span of these mice suggesting that underfeeding has an important effect on reducing autoimmunity. This finding, of course, also lends further support to the immunological theory of aging.

CRITICISMS OF FOOD RESTRICTION RESEARCH

Although most researchers agree that dietary restriction provides an important means of understanding the causes of aging, there is nevertheless a body of opinion which believes that all calorie restriction experiments are fundamentally flawed. The main reason for this lies not so much with the food restricted animals themselves, but with those with which they are compared, namely the fully fed controls. It has been pointed out that in the "real" world, unlimited food supply rarely occurs, and consequently most wild animals are subjected to intermittent feeding and are thus naturally calorie restricted. Therefore, the increase in life span of restricted animals may be more apparent than real since it may be a result of fully fed animals dying before their time. Indeed, considering that experimental animals are often confined to a relatively small cage, without the means for vigorous exercise, and with eating as one of their few pastimes, then it can be argued that this situation is hardly a normal one. In fact, many humans in the

Western world may be leading an analogous life – being fully-fed (and often overweight), and coming home from the office to sit in front of the television without taking part in much exercise. And as most doctors will testify (see below) this type of situation, especially in middle age, can provide a recipe for premature death. Although the use of fully-fed control animals is far from ideal, it is however, almost impossible to design food restriction experiments any other way. The only means of overcoming these criticisms would be to perform studies on wild animals; but of course, it would be impossible to control the quantity and quality of the diet, and furthermore, such animals would probably be eaten by predators before they reached the end of their life making the study invalid.

WHAT IS THE RELEVANCE OF FOOD RESTRICTION TO HUMANS?

Although the modern study of food restriction on longevity starts with the work of McCay it can be argued that human beings have actually known about the benefits of food restriction for much longer. For example, the Renaissance autobiographist Luigi Cornaro (1464–1567) wrote four volumes on how to increase life span entitled *Discourse on the Sober Life* in which he advocated dietary restriction as the means of extending life. Conaro was a member of the Italian nobility and, for the first 37 years of his life, by his own admission, led an excessively gluttonous life which resulted in him suffering dangerous ill health. However, at this point he changed his ways and adopted a restrictive diet that he followed for the rest of his life, eating only small portions of bread, meat, broth with egg and good wine. This regime worked since Conaro lived to be 103, and remained in good health up to the end of his life.

Despite this, for obvious ethical reasons, it is not possible to repeat the severe dietary restriction undertaken on laboratory animals with humans. For example, to achieve the longest increase of life span, animals probably need to be put on restricted diets after weaning and before they become adults. To do the same with humans, infants would have to be systematically starved to such an extent that they would probably never reach full adult size. And the payoff, if any, would not be known for at least 70 years or so. Clearly this type of study is out of the question. However, such work can be undertaken with monkeys and two such studies are now underway. The first led by George Roth of the National Institute of Aging was begun in 1987, and is examining rhesus monkeys (which typically live up to about 30 years), and squirrel monkeys (which rarely live beyond 20 years). These animals have been placed on a level of calorie restriction which is about 30% below normal, with some monkeys beginning their dietary restriction in youth (at one to two years) whereas others were started at puberty. The second study initiated in 1989 at Wisconsin-Madison University is using a similar protocol except that calorie restriction is to be implemented when the monkeys become young adults (eight to fourteen years). According to Richard Weindruch in a recent (1996) *Scientific American* article, the preliminary results are encouraging. The dieting animals appear healthy and happy, although eager to be fed. In addition, their blood pressure, insulin and glucose levels are lower than fully-fed animals. Of course it is going to be many more years before the full results of this study are published, but it should provide us with the best evidence yet concerning the relevance of dietary restriction for human aging.

Despite the lack of human evidence, researchers have nevertheless attempted to provide guidelines for those interested in following a restricted dietary regime. For example, Richard Weindruch believes that many people would do best to regularly consume an amount of food that enabled them to weigh 10 to 25% below their own personal set weight (that is the weight which appears to be most 'natural' for them). Weindruch also recommends reducing one's daily intake of protein to roughly one gram for each kilogram (2.2 pounds) of body weight, and reduce fat even further to about half a gram per kilogram. Of course, such a dietary regime has not been experimentally proven to work in humans – but all things considered, one might nevertheless expect to gain some benefit from following such advice.

DIETARY PRACTICES IN OTHER SOCIETIES

There are many societies, particularly in the third world, where large numbers of people suffer from calorie restriction. However, such people are also typically undernourished in terms of vitamins and minerals and, in the quality of food they consume. Thus, they obtain no benefits of calorie restriction. But, there are some important exceptions, and perhaps the best known are a population of people who live on the Japanese Island of Okinowa (see previous chapter). Accurate birth records of these people have been kept for over 100 years, and these show that Okinowa has 2 to 40 times more centenarians than any other part of Japan (and this is despite the fact that Japan has more centenarians than any other country in the world). The Japanese authorities, however, not only keep accurate birth records; but since 1946 they have also carried out annual surveys on nutritional intake throughout Japan which involve personal interviews and the weighing of food eaten during a period of three consecutive days. The results show that Okinowans differ greatly in terms of their diet compared to other Japanese. For example, the intake of sugars is about 25% of the average for Japan and the intake of cereals around 75%. Consequently, the total energy consumed by Okinowans (particularly in early life) is only 62% of the recommended intake for Japan. Despite this, Okinowans eat far greater amounts of green-yellow vegetables (300% of the Japanese average) and fish (200%). In short, Okinowans eat a restricted but healthy diet. Although the average weights and heights are less than on mainland Japan, the Okinowans are generally more healthy and suffer less cancer, heart disease and stroke.

Okinowa is not the only place where people appear to live longer than might be expected by chance. Geronologists have long known of three other places in the world where people claim to live extraordinary ages (often beyond a hundred years) whilst enjoying excellent health. These places are the tiny isolated village of Vilcabamba in the Andean mountains of South America; the district known as Abkhazia in the Caucasus Mountains of Russia; and the province of Hunza in Pakistani-controlled Kashmir (see also page 175). These remote communities have received international attention and scientific visits in recent years and, in general, most scientists have been impressed with the claims of their inhabitants; although it must be said that there is very little solid documentary evidence (i.e., no reliable birth records) to substantiate the ages. One scientist who has visited all three regions in an attempt to find a common denominator is the distinguished American physician Alexander Leaf who published his findings in *National Geographic,* as well as writing a longer account in his book *Youth in Old Age* published in 1975.

Although Leaf found a number of differences between the three communities, he nevertheless showed that there were certain lifestyle patterns that appeared to be common to all – one of the most obvious being diet and calorie consumption. All three groups ate mainly vegetables (only the Abkhazians ate any appreciable amount of meat or dairy produce) and Leaf estimated that the daily calorie consumption of the three regions ranged from around 1,200 to 2,000 calories. In short, they all consumed a diet that was significantly less in calorie intake than those normally eaten in the West. But perhaps even more impressive was the high level of physical activity common to all three regions. All three were mountainous locations (ranging from about 3,000 to

7,000 feet) and, as Leaf pointed out, a substantial amount of physical exertion was necessary just to undertake the daily business of living. Moreover, there was no retirement, with people working throughout their lives, and forming an important and respected part of the social fabric of the community. Despite this, all three groups were not without their vices. Smoking was surprisingly common in all three regions and, in addition, all three communities brewed their own potent alcoholic concoctions – in fact, these alcoholic brews were often perceived by the longest lived individuals to be the most important determinant of their longevity!

The work by Alexander Leaf indicates that there are several factors that underlie increased longevity, but clearly a healthy and moderate diet is one of them. In fact, there is even some evidence that restricted diets in Western societies can have beneficial effects upon the health of the population. For example, during the Second World War the population of most European countries were forced to survive on rationed amounts of food. Although calorie restriction was often severe, attempts were generally made to maintain adequate levels of vitamins. Surprisingly, examination of mortality statistics from Sweden, Finland, and Norway for the 1940s suggest that the state of health in these nations was never as good and, in particular, there appeared to be a significant reduction in mortality from heart disease. However, after the war with the removal of food restrictions, and increased calorie intake, the mortality from heart disease soon increased back to pre-war levels.

BODY SIZE AND HEALTH

So far the evidence would appear to be overwhelming in support of the idea that being slim and underweight is beneficial both in terms of health and longevity. However, not all the evidence supports this conclusion. For example, Reuben Andres, working at the National Institute of Aging in Baltimore, has examined life insurance statistics to assess the relationship between mortality and body weight. From what has been said so far, it might be predicted that thinner people will always live the longer. However, this is not what Reuben Andres found: in contrast to expectation, he showed that a U-shaped curve best fitted the relationship, with both the thinnest and the fattest people showing the highest mortality. In fact, even more damaging to the food restriction theory was the finding that the people who lived the longest were those who were actually 10 to 25% *above* ideal body weight. These findings have been re-examined and confirmed by life insurance companies, and has led to a revision of height-weight tables on which their premiums are partly based.

Perhaps, not surprisingly, there have also been criticism of Andre's interpretation of his data. For example, it has been pointed out that the proportion of smokers were far greater in his underweight group (80%) than in the overweight group (55%). Furthermore, others have suggested that the underweight group may have had greater rates of subclinical disease (i.e., disease that has not yet been diagnosed). In fact, some researchers (Manson *et al.*, 1987) claim that when these biases are taken into account then minimum human mortality occurs at weights at least 10% below the average. It is difficult to gauge who is correct and, consequently, a large deal of controversy surrounds this area (several large scale surveys have also provided inconclusive evidence on this subject). It has also been pointed by Richard Weindruch and Roy Walford, that even if Reuben Andres is correct, his data only examines body size and *not* calorie intake and, therefore, does not disprove the evidence that has built up showing the effectiveness of dietary restriction.

Despite the uncertainty regarding the ideal weight to which we should be striving for, there is little doubt that being significantly overweight, or obese, has serious health risks. One of the most important studies bearing on this issue is the Framingham study; a study set up by the US government in 1948, in an attempt to discover the most important causes of heart disease. This study has followed 5,209 men and women from the town of Framingham (located 21 miles west of Boston) with regular health check-ups taking place every 2 years. The results of this study have identified a number of risk factors that increase the chances of heart disease, and one of the most important is obesity. For example, the Framingham study has shown that high blood pressure developed 10 times more frequently in people who were 20% or more overweight. In turn, the risk of heart disease in those with hypertension was 2 to 3 times greater than normal, and the risk of stroke was even greater with hypertension increasing its likelihood seven-fold. Other studies derived from insurance statistics also confirm these findings. For example, Dennis Bellamy in his book *Aging: A Biomedical Perspective* has pointed-ed out that between the ages of 20 and 64 years, obesity increases the chances of mortality by 50%. Furthermore, not only is heart disease increased but there is a 90% increase in kidney disease and 400% increase in diabetes. In addition, cancer incidence is also higher with one set of figures revealing that mortality from cancer in people of normal weight was 111 per 100,000 which increased to about 140 in people who were 15–25% overweight. In short, if one is overweight there is perhaps no better advice than to go on a diet (preferably under medical supervision).

SUMMARY

- Calorie restriction, or reduced feeding without malnutrition, has been shown to increase maximum life span in laboratory animals by up to 50%.
- Although calorie restricted animals tend to be smaller than normal, they are fitter and leaner, and suffer significantly fewer diseases. Most biological measurements (although not all) show that these animals are 'younger' than normally fed animals.
- The greatest increase in life span is obtained if calorie restriction is started early in life (and preferably just after weaning). However, significant life extension can also be achieved if food restriction is started later in life, or even if it is intermittent (i.e., one day in three)
- The important factor in food restriction would appear to be the number of calories consumed and not any change in the protein or fat content of the diet itself.
- Calorie restriction cannot be explained on account that it slows down growth, or because it reduces body stores of fat.
- Despite the plausibility of the rate of living theory to explain the effects of calorie restriction (i.e., it works because it slows down metabolism) recent evidence does not support this idea. Food restricted animals appear to have normal levels of metabolism although there may be differences in the way these animals utilise their energy.
- Evidence shows that calorie restriction reduces the number of free radical reactions taking place in the cell, and also helps to provide a better free radical defence by increasing levels of antioxidant enzymes such as superoxide dismutase.
- Calorie restriction depresses the release of many hormones in the body (it may do this in part by acting on the hypothalamus) and it also slows down the age-related decline of the immune system (perhaps by its effect on the thymus gland).
- Some researchers believe that the conclusions derived from calorie restriction research are flawed because laboratory conditions allowing animals unlimited access to food are 'unnatural' – and thus they cannot provide meaningful comparison with restricted animals.

- The Framingham study which followed a total of over 5,000 men and women over a number of years found that being overweight was a major risk factor for several life-threatening diseases. However, the ideal body weight for health and longevity still remains a matter of controversy.

FUTURE PROSPECTS

*One fact about the future of which we can be certain is
that it will be utterly fantastic*

–Arthur C. Clarke

*We used to think our future was in the stars.
Now we know its in our genes.*

–James Watson

The main aim of this book has been to track down the fundamental causes of aging – or at least, point the reader to its most probable causes. Underlying all the research and arguments is the assumption that there is, and can be, nothing intrinsically unfathomable about the aging process. Aging must logically be the result of various biochemical and physiological changes that take place in our body as we get older, and the causes of this change must follow clearly definable biological and scientific laws. Although aging appears to be inevitable, and because our own eyes seem to tell us there is little we can do to halt its progress, it is tempting to believe that aging is either inexplicable or too complex to be understood. But hopefully, as this book has shown, this view is unduly pessimistic. There is no good scientific reason why the *exact* causes of aging cannot be fully explained and, almost certainly, it is only a matter of time before this occurs. But, of course, this is only one side of the coin. With knowledge comes the increased likelihood that one day meaningful intervention into the aging process, will be possible with the promise of a greatly extended life span, or even immortality. Moreover, an understanding of aging will almost certainly allow greater control and prevention of the diseases that accompany old age, and perhaps even disease in general. Indeed, few would deny that the conquering of aging has the potential to transform the world as we know it, and to be of benefit to all humankind. But is this a realistic possibility in the coming years? Can we really ever hope to understand aging? And, if so, can we really stop or slow down the process? Of course, nobody can look into the crystal ball and see the future, and any attempt to do so is bound to be proven wrong, if not ridiculed, within a short space of time. However, one can at least attempt to make some educated guesses and this is what this chapter sets out to do.

THE HISTORY OF SCIENTIFIC PROGRESS

It is easy to forget that most of our scientific and medical knowledge is primarily a development of this century, if not the last half century. For example, many of the technologies we now take for granted including automobiles, aeroplanes, telecommunications, nuclear energy, space travel, computers and oral contraceptives, to name but a few, have all been developed this century, and within the span of a single human life. And, if anything, we tend to underestimate what the future may bring. To give an example: in 1937 the National Resources Committee reported to the American President, Franklyn D. Roosevelt, about probable and important future technological developments, and totally overlooked the discovery of nuclear power, jet propulsion, and the transistor – all of which were in use within 15 years of the report and have since transformed the world. As this example shows, it is difficult to predict the future, and it is not getting any easier to judge the limits of what can be achieved as scientific progress speeds up. In fact, it has been estimated that every 10 years our total scientific knowledge is doubling. Consequently, in 1950, humans had access to twice the amount of knowledge about the world than they had in 1940. By 1960, scientific knowledge had expanded by four times compared to 1940; by 1970 eight times; by 1980 sixteen times; and by 1990 over thirty times. By the end of the millennium we will have approximately 64 times the amount of knowledge than we did in 1940. And, accompanying this knowledge will be scientific and medical breakthroughs that will continually change the balance of what can and cannot be done. It takes no great insight to realise that if these trends continue then the next century will undoubtedly be a time of great discovery and breakthrough.

Scientific research into aging is no exception to these general trends. For example, it wasn't until 1945 that the first scientific journal dedicated to the subject of aging (the *Journal of Gerontology*) began to be published, and later developments included *Experimental Gerontology* (started in 1964), and *Mechanisms of Aging and Development* (started in 1972). Today there are over 35 periodicals that are devoted to aging, proving that we are increasingly learning more about the subject. Despite this, only a few of these journals actually publish findings about aging from a biological perspective. Indeed, it is a rather surprising fact that aging is still very much a minority discipline in biology, and perhaps even more surprising that the subject of biological aging continues to be a minor interest area in gerontology. Thus, despite the scientific explosion of interest into aging over the last few decades, research into the biological causes of aging has tended to remain in the background. It is not even clear if knowledge in biological gerontology is expanding at a rate consummate with other areas of gerontology, and even if it is, its still the case that other areas of aging research tend to be better funded and supported. Thus, as the discipline of gerontology rolls into the 21st Century, it would appear that research into the biological causes of aging is not exactly at the vanguard of this advance.

One of the most important developments in the field of gerontology in recent years has been the founding by the United States government of the National Institute of Aging (NIA) in 1976, which makes a firm commitment to funding research in aging. Indeed, the NIA is now recognised as the premier research institute on aging in the world, and in 1993 it had a budget of over 400 million dollars, with most of this money being spent on research. Despite this, the NIA is responsible not only for research into the biological and medical aspects of aging, but for the social and behavioural aspects as well. Consequently relatively little research funding actually gets spent on attempting to understand the basic *causes* of aging. For example, about half of the NIA budget is now spent on Alzheimer's disease and other diseases common in the aged, and another substantial part of the budget is spent on incontinence, falls, psychiatric, psychological and social problems. Indeed, as Leonard Hayflick has pointed out in his book *How and Why We Age*, it is probable that less than 50 million dollars is presently spent on research into the basic causes of aging. To put this into perspective, this is about the price of television advertising during the 1993 Superbowl, and is a very small amount compared to the level of funding for other areas of medical research such as cancer and heart disease. In short, research into the biological basis of aging, while ticking away slowly, certainly does not enjoy the limelight and does not receive the support it arguably deserves.

One can only guess at the probable breakthroughs and benefits if there was a significant increase in funding, either in the US or world-wide, into the fundamental causes of aging. It is sometimes said that aging is so complex that it will never be fully understood. But, by the same token, it can equally be argued that considering the relatively small amount of money that has been spent on this type of research, and the small band of scientists involved in this area, then it is perhaps surprising that we know as much as we do! If we adopt the view that we still have much to learn (and achieve), then it is clear there are grounds for considerable optimism about future progress in understanding the causes of aging.

Despite this, there appears to be little prospect that research funding or academic interest into the biological basis of aging is going to dramatically increase in the immediate future. Does this mean, therefore, that progress is going to be slow and unspectacular? The answer to this question is almost certainly no. The next few decades are likely to see major breakthroughs in understanding aging, although the answers may not necessarily come from aging research, but from other areas of biology. Perhaps the most exciting and important of these developments is the setting-up of the human genome project. Although this endeavour is not currently regarded as part of gerontological research, it is probable that if the human genome project fulfils its promise, then at some point in the future the disciplines of genetic biology and gerontology will come together and bring with it a brand new dawn of aging research.

THE HUMAN GENOME PROJECT

The aim of the Human Genome Project is simple: it is to determine the complete sequence of the 3 thousand million base pairs, or so, that make up our 23 pairs of chromosomes (the human genome) and, to identify amongst these random bases the 100 thousand or so genes that uniquely define us as human beings. The notion of the human genome project had its beginnings in the United States in the mid 1980s, and was put into practice in 1990 as an internationally organised research effort involving hundreds of laboratories around the world. In addition to deciphering the human genome, the project also involves the complete mapping of the genome of several other organisms that are also important to genetic research including the bacterium (*Escherichia coli*); the yeast (*Saccharomyces cerevisiae*); the nematode (*Caenorhabditis elegans*); the fruit fly (*Drosophilia melanogaster*); and the laboratory mouse (*Mus musculus*). Part of the reason for this extra work lies with the fact that many species share similar genes with our own. For example, genes involved in human cancer have been found to be similar to certain genes in worms where they can be more easily studied. Furthermore, comparing genes between species also helps researchers understand how genes have evolved thus allowing further insight into their function.

However, the mapping of the human genome is an arduous and mammoth task. As we have seen in previous chapters, our DNA is made up of only 4 bases (A,C,G and T), and because there are two strands to each DNA molecule, these bases are therefore arranged in pairs. It is the aim of the human genome project to list these bases in their correct sequence. The task has been called the holy grail of modern biology, and it is arguably the most ambitious human endeavour that has taken place since the Apollo space program landed a man on the moon. At the beginning of the project it was estimated that it would take some 30,000 man years of laboratory work at the cost of some $3 billion to complete. While new technological developments in sequencing DNA has cut down the laboratory hours, it is probably the case that the total cost has significantly increased. Despite this, work on the project has already resulted in a rough map of nearly all of our chromosomes, and it is estimated that the human genome should be finished before the year 2006. The reward will be, in effect, to fill the equivalent of 13 sets of the Encyclopaedia Britannica with rows upon rows of letters representing our 4 bases. As Enzo Russo and David Cove have written in their book *Genetic Engineering: Dreams and Nightmares*, if we assume that an average book has around 3,000 letters per page, we would need a million pages to write the whole list of our bases, or rather 5,000 books of around 200 pages each. In terms of bedtime reading it would undoubtedly be the most boring book ever written (not that it will ever be published in this way), but as a potential set of instructions for understanding the biological nature of human beings it will have no equal. It is certain that this knowledge, with its many ramifications, will dominate human biology and medicine over the next millennium.

Why then are researchers going to such trouble to map the human genome? There are several good reasons. For example, sequencing the structure of genes will enable their identification, and one of the first benefits of this should be that it will allow doctors to identify the most important genes that contribute to disease. Indeed, it is feasible that in 20 years time it will be possible to take the DNA from a new-born infant and analyse 50 or more genes that are known to predispose us to many common diseases including heart disease, cancer and dementia. Consequently, from this genetic information, a lifestyle and medical regime could be drawn up that will help the individual minimise their chances of disease, and maximise their chances of a long healthy life. Despite this, being able to identify genes in this way actually tells us very little about their function; and in the long term, the great advantage of the human genome project will not only be in the identification of genes, but also in establishing what proteins they make, when they act, and the functions they serve. This is a huge task and the accomplishment of the human genome project will only be the starting point for this much more complex and enlightening journey which will require vast resources (and money), and the involvement of huge numbers of scientists over a long and extended period of time. But the rewards are likely to be great – in short, the total understanding of how genes create the body; how they control and maintain it throughout life; and how they lead to its demise through disease and aging. And, at every point along the way, scientists will probably have increasing power to intervene, and to tinker with the very basic processes of life itself (see also pages 290–295).

THE IMPLICATIONS OF THE HUMAN GENOME PROJECT FOR AGING

As we have seen in this book, there are still many basic questions to be answered about the role of genes in aging. But perhaps the most central of all, is whether aging is a genetically programed process or a result of random wear and tear. This is a simple and fundamental question which still awaits a satisfactory answer. Indeed, most of the theories discussed in this book, whether one looks at genetic, cellular or organ changes, can be explained from either perspective. In short, we still do not know if age-related changes are designed to occur, or whether they are the result of the body simply wearing itself out. The problem is all the more acute because, until this basic dilemma is solved, research into the understanding the causes of aging is certain to be difficult and lack direction.

How then might the human genome project help contribute towards solving this puzzle? One way might come from the impetus that it will give to understanding human development. A central issue in biology, and one that has many implications for medicine, is the question of how the human genome can build the body from almost nothing; and along the way transform itself into an embryo, a new-born infant, and then eventually into an adult. There is still much to understand about how the fertilised egg develops into the complex array of specialised cells that constitutes the body, although most biologists are confident that the human genome project will allow the process to be explained in considerable depth – perhaps to the extent that it will be possible to map out the whole sequence of development from egg to adult in one large flow diagram. But where does this programed development end? Hopefully, this is a question that the human genome project will help solve and in doing so, answer once and for all the basic issue of whether aging (or old age) is a genetically programed process, or not.

Although there is no certainty that aging genes will be found to exist, suppose researchers discovered that certain aspects of aging were indeed genetically programed to occur at a certain time. What use could this information possibly have? The first thing to bear in mind is that genes do not work in isolation: they contain structural components (called operons) that can be likened to on–off switches (see page 147), and they need special chemicals to activate them into transcription. In some cases these genetic switches are known to be activated by hormones, and in other instances they are turned on and off by substances known as gene regulatory proteins. By understanding the chemical mechanisms that control genetic activity, it may be possible one day to design drugs that can be injected into a person that target these genetic on–off switches to restore youthful or genetically controlled functions. Alternatively, it may be possible to simply inject the person with new genes. As we shall see below, there are grounds for believing that this will become a distinct possibility in future years.

On a less ambitious scale the human genome project should also allow the genetic make-up of exceptionally aged individuals to be examined to see whether they share any genetic similarities. We have already mentioned a study of this sort in chapter 10, which showed that the long lived people on the island of Okinowa tended to lack a certain gene (called DRw9). However, with the means of being able to detect *all* the 100,000, or so, human genes, it should become possible to associate human longevity with many more DNA regions, and their gene products, thus giving scientists a further powerful research tool to understand aging.

Even if aging is not programed to unfold, it still remains that genes are certainly involved in the aging process. For example, some researchers such as the late George Sacher believed in the existence of longevity assurance genes (see chapter 2) and evidence suggests that these genes may make up a relatively small proportion of the human genome (perhaps no more than 250 genes). If this is so, then the human genome project may well eventually lead to their detection and, more importantly, an appreciation of how they work. Alternatively, a number of evolutionary biologists believe that aging is not so much due to the protection provided by longevity genes, but is the result of antagonistic pleitrophy where a build up of harmful genes (or rather their substances) occur towards the end of life. Again, it may be that in the long term the human genome project will open up ways of proving which theory is correct, and on the basis of this knowledge offer new approaches to intervention. For example, if pleotrophic genes underlie aging, then it should be possible to discover the harmful substances that they are producing and a way of minimising their effects by developing pharmacological treatments.

GENETIC ENGINEERING

The human genome project, however, is only one of the exciting developments taking place at present. Another important advance is genetic engineering; that is, the ability to take genes from one organism and to transplant them into another. Although farmers have selected plants and animals for breeding purposes for over 10,000 years and, in effect, produced genetically engineered life forms; it is only recently that scientists have been able to make copies (clones) of genes and to introduce them directly into the genome of another organism. In fact, genetic engineering is already taking place on a large scale. It has been used in farming to increase crop yields and to make plants more resistant to disease, and some animals have been given genetically engineered products to make then grow bigger, or increase meat and milk yields. In addition, genetically engineered cells have been used commercially to mass produce hormones, antibodies and vaccines; with a million dollar biotechnology industry springing up to take advantage of these developments. Genetically engineered cells have even been introduced into human beings to correct a number of genetic diseases (see below) although such procedures are very much in their infancy.

The beginning of genetic engineering, or perhaps more accurately the start of what is known as DNA recombinant technology (recombinant DNA is constructed outside the living cell by splicing 2 or more pieces of DNA from different sources to provide a novel combination of genes not normally found in nature) took place in the early 1970s when geneticists discovered a class of enzymes (called restriction enzymes) that were able to cut DNA at a specific base sequence and with great precision. These enzymes were first found in bacteria who used them to disable invading viruses (that is, if a bacterial restriction enzyme came across viral DNA it would sever it into pieces). But it didn't take long for researchers to realise that restric-tion enzymes could also be used in the laboratory as molecular scissors to cut DNA. In short, DNA could be extracted from a living organism, isolated, put in a test tube with restriction enzymes, and then a little while later lots of DNA fragments would be obtained that were all neatly snipped at the same place. In fact, within a decade of their discovery, geneticists had discovered over 300 enzymes, from a variety of bacteria, that enabled them to cut DNA at many different places.

By themselves, lots of DNA snippets floating around in a test tube have little use. However, a couple of years after the first restriction enzymes were found, geneticists also discovered that some of these enzymes also produced DNA fragments that had 'sticky ends', which allowed the snippets of DNA to be joined together. In fact, for this to occur, the DNA didn't even have to come from the same genome. In other words, it was possible to join bits of DNA together from different organisms – or in effect, to implant new genes into foreign DNA. But, again, all of this recombination was taking place *in vitro*, that is outside the cell (or in the confines of a test tube). The important step was to find a way of introducing new DNA into living cells, and this was first achieved by Paul Berg and his colleagues at Stanford University in 1972, who managed to cut genes from bacteria and then to splice them into a virus called SV40. A year later came an even more astonishing feat. This time, Herbert Boyer and Stanley Cohen (again of Stanford University) took the DNA from the African clawed toad and placed it inside the genetic material of the *Escherichia coli* bacterium. Even more remarkable was the fact that every time the bacteria divided, the toad's DNA was also copied, and passed into the new bacteria. Thus, the new foreign gene had not only been incorporated into a new host – but it was also there to stay for all future generations.

The technique of artificially inserting genes into rapidly dividing organisms such as bacteria was to have a huge impact on research, because it allowed for the first time a means of cloning, or making large numbers of identical genes. In effect, bacteria modified in this way could be likened to factories that were able to make endless copies of its implanted genes. Because bacteria are able to divide approximately every 20 minutes, a single gene placed into a bacterial host could therefore be multiplied many times in a short period of time (in fact, if given adequate nutrition, bacteria could in theory produce a mass greater than that of our planet in less than 2 days!). Thus, bacteria quickly became the assembly lines by which recombinant genes could be made. This new technique revolutionised the study of molecular biology because for the first time it allowed large amounts of new DNA to be produced, thus enabling investigators to work out, with relative ease, the sequences of bases making up genes. This was important because once a gene was cloned and sequenced, it was then possible to predict the type of protein it made, and then to work out what the protein might do. Thus, by the middle of the 1970s, armed with the tools of being able to cut, paste and clone genes, researchers had at their disposal the resources to understand and manipulate the genetic basis of life.

MEDICAL USES OF GENETIC ENGINEERING

Perhaps the most exciting prospect for genetic engineering is gene therapy – that is, the treatment of disease by introducing healthy genes into the body, particularly to replace those that are faulty through genetic disease. There are at least 2,000 different diseases that are caused by a defect in one or more genes, and most lack fully effective treatments. In theory, at least, gene therapy offers the possibility of being able to correct all these disorders by introducing a copy of the normal gene, or genes, into the appropriate cells of the body. In fact, gene therapy in humans was first carried out in 1990, at the National Institute of Health in Bethesda, USA, to treat a disorder called Severe Combined Immune Deficiency in a 4 year old girl called Ashanti de Silva. This inherited disorder is due to a defective gene found on chromosome 20, which results in a deficiency of an enzyme called adenosine deminase, that is vital to the production and maintenance of white blood cells used to fight infections. As a result, affected children are unable to defend themselves against infections and have to live in a germ free environment. This makes keeping them alive a daunting prospect. Even then, these children rarely live to more than a few years of age.

It might be thought that the simplest approach to treatment would be to replenish or inject the missing adenosine deminase back into the body. Unfortunately, adenosine deminase only exists in the body for a few minutes before it is broken down and therefore this strategy does not work. However, because the gene responsible for making adenosine deminase had been identified and cloned by the early 1980s, doctors had all the pieces in place to correct the disorder through genetic engineering. And, in September of 1990, this type of operation first took place. The procedure was relatively straightforward. To begin, the young girl's white blood cells were removed by a technique called apheresis (where blood was drawn from an intravenous needle and white blood cells extracted) and then put into a culture which contained a harmless mouse virus that had been engineered to contain the normal adenosine deminase gene (a virus is simply a small parcel of DNA that can not replicate by itself and therefore has to embed its genes into the DNA of its host to help it out). The virus carrying the adenosine deminase gene had been disabled (so that it could not replicate itself) and was then allowed to infect the culture – thereby inserting its 'modified' DNA (including the adenosine deminase gene) into the white blood cells. Ten days later a billion of these 'infected' white blood cells were infused back into the patient in a simple transfusion that took 28 minutes.

It wasn't long before investigators began to see the benefits of the procedure. Over the next 6 months, along with two more infusions, the patient's level of adenosine deminase increased and the number of her infections began to decline significantly. In fact, she is now able to lead a relatively normal life, attend school, and moreover she appears no more likely to catch infections than her classmates. Indeed, on one occasion when all her family caught flu, she was the first to recover. There is no doubt that the procedure has been a great success, although the operation is not a permanent cure since white blood cells live less than a year, and they constantly need replacing. Consequently, Ashanti will require regular transfusions every few months of genetically engineered cells for the rest of her life. Nevertheless, most will accept that this is a small price to pay for the existence of normal and healthy life.

There are still many obstacles to be overcome before gene therapy becomes commonplace in hospital operations. For example, to begin with, the gene responsible for causing the disease must be isolated and cloned in a laboratory. More problematical is the fact that for the majority of conditions, the gene must be inserted into the right tissues. This was relatively easy to do for adenosine deminase deficiency because the problem was with the white blood cells which could be easily removed from the body. However, for other diseases the situation is not so straightforward. For example, gene therapy of cystic fibrosis would require the normal gene to be placed into the cells of the persons lungs, and in the case of Alzheimer's disease the genes would have to be placed into the brain. Even then, to function inside a living cell, the gene will need to be integrated into the chromosome and contain the correct promoter sequences to allow it to be transcribed from DNA to RNA. Moreover, because researchers have little control over where the new gene is to be inserted, there are fears that if positioned in the wrong place it may activate a cancer causing oncogene or inactivate a tumour suppressor gene. Thus, there are many biological (and indeed ethical) problems to be solved. Nevertheless, by 1995, over 200 trials for gene therapy had been approved world-wide, to treat a wide range of conditions including cystic fibrosis, Duchenne muscular dystrophy, haemophilia, thalassaemia, Parkinson's disease and a variety of cancers.

THE IMPLICATIONS OF GENETIC ENGINEERING FOR AGING

It is certain that experiments involving genetic engineering will provide an important research tool by which to understand aging in the coming years. In fact, such work has already started. For example, William Orr and Rajinda Sohal (as discussed in chapter 3) have genetically engineered fruit flies to carry extra copies of genes that encode for the antioxidant enzymes superoxide dismutase and catalase, and found that these flies gained a significant (one-third) extension of life span. However, instead of targeting a specific type of cell (such as white blood cells) these investigators placed the antioxidant genes directly into the embryo – with the consequence that practically every cell of the adult fly carried the extra genes. Thus, the antioxidant enzymes provided extra protection throughout the body. Of course, this form of 'germ cell' engineering has not been undertaken with humans; although it is increasingly being used in laboratory animals, where it has become a powerful experimental tool in many areas of biological research. However, at present its use in aging research is limited because relatively few genes have been identified (and cloned) that are known to be directly linked to aging. Despite this, one feels that it is only a matter of time before genes from a wide range of species including yeast, nematodes, fruit flies, mice, and perhaps even humans (these could be transplanted into other animals to assess their function) are used in similar ways to explore their role in aging.

As an experimental tool, the development of transgenic animals is likely to provide many valuable insights into understanding the genetic basis of the aging process. But, in the long term, the benefits of this work may even extend to direct intervention into the aging process. There are many possible ways this could happen. For example, as we have seen, fruit flies live longer if given extra antioxidant genes. But what would the consequences be if similar genes were given to higher animals such as rodents and primates? And if this genetic engineering of longevity worked for these animals, would humans seeking to extend their life span be allowed to follow the same path? It is interesting to ponder such questions because if more studies are published that support the notion that life span can be significantly extended by the insertion of new genes into higher animals, then the question of similar human engineering may well achieve greater prominence (and no doubt heated debate) at some point in the distant future.

There are many other potential ways in which humans may eventually use the technique of genetic engineering to extend their life span. For example, on a genetic level it may be possible to engineer more effective (or youthful) DNA repair systems: or on an organ level it may be possible to engineer a better immune system. Advances on both these fronts would be likely, for example, to reduce the risk of developing cancer and increase longevity. Alternatively, on a cellular basis, if the telomere theory is shown to be correct, then it might be possible to remove certain cells from the body and to culture them so that their telomeres can be artificially lengthened and therefore rejuvenated. Of course, this is at present all science fiction, and it is easy to dream up many other far-fetched ideas on the same theme. Nevertheless, it is certain that we stand on the threshold of a brand new dawn with regards to our ability to genetically manipulate living organisms; and it may be that in the long term, genetic engineering will offer the greatest hope of all, for those of us who want to extend life span beyond its natural limits.

WEAR AND TEAR APPROACHES

It is easy to lose sight of the fact that despite all the exciting advances taking place in genetic engineering and the human genome project, aging may not be a predominantly genetic process! Although, as we saw in chapter 3, longevity is inherited to some extent, the effect of this inheritance does not appear to be very strong. For example, if one looks at the longevity of identical twins, one finds that they are more likely to have similar life spans than non identical twins. But, the correspondence is rarely exact, and research shows that, on average, identical twins will die 3 (or more) years apart. Similarly, if one has long-lived parents, the chances are that one will live longer than average, but nevertheless not as long as one might hope to expect. Although it is difficult to put an exact figure on to what extent longevity is due to genetic and lifestyle factors, it is probably the case that for most people lifestyle factors are the more important. Thus, working out how genes are involved in the aging process may only provide an incomplete picture of how aging proceeds. Similarly, future genetic intervention into the aging process may perhaps only provide a partial extension of life span.

The main alternative to the genetic explanation is the theory that aging is the result of accumulating damage to the tissues of the body as a result of wear and tear. There are indeed many potential sites where such damage is likely to take place. For example, at the molecular level, cross-linkages may occur within and between molecules; DNA bases become damaged; and errors made in the transcription and translation of genetic information. There are also many biochemicals that have the potential to cause molecular damage including glucose (and its break down products) and elements such as sulphur. However, by far the most important class of agents believed to inflict damage on biological systems are free radicals derived from oxygen, or other by-products of fuel combustion in cells. Free radicals are truly ubiquitous and cause damage to DNA, proteins, and to the many other components of our cells including membranes and mitochondria. Indeed, as mentioned in chapter 5, it has been estimated that each and every one of our cells receives over 100,000 oxidative 'hits' every day with the mitochondria taking the brunt of the attack. It is hardly surprising, therefore, that many investigators believe free radicals to be the ultimate cause of aging and biological decline.

However, it is difficult to see how any future form of genetic engineering could ever provide complete protection against the effects of free radicals. Of course, being engineered with increased levels of antioxidant enzymes may slow down the rate at which free radical damage proceeds, but with such a continuous and overwhelming bombardment of oxidative hits, then such damage is always bound to occur and accumulate. Perhaps one day in the future scientists will be more successful in preventing free radical damage by the development of more powerful antioxidant drugs. But, again, it is hard to imagine how a drug could be developed which offers complete protection against the effects of free radicals. Since they are always going to occur, one can only hope to minimise the damage. Indeed, it must be said that scientists already have powerful antioxidant drugs and have given them to laboratory animals in huge doses; and the result is that they help delay the onset of disease but do not increase their maximum life span. In short, the prospects do not look good for developing a drug that will stop all free radical damage from taking place. But, unfortunately, as long as free radical damage continues – probably so will aging.

NANOTECHNOLOGY

If aging cannot be totally prevented by blocking the effects of free radicals, then what are the chances of being unable to undo the damage? Of course, one of the defining and most frustrating characteristics of aging is that it is irreversible. But what if we could develop repair systems that were able to put right damage that has already occurred? To do this we would probably have to go down beyond the molecular level – to the atomic level – in order to clean up all the atomic debris and to replace all the particles that have become displaced and missing. But what if we really could go down into the atomic depths? In fact, some scientists believe that one day this may become possible with the development of nanotechology: in short, the technology of miniaturisation which is able to manipulate and control things on an extremely small scale. The idea of nanotechnology is usually credited to the physicist Richard Feynmann, although the concept has been more recently developed and popularised by the work of Eric Drexler. The idea is simple enough: to build tiny machines or computers that can literally move atom by atom. At present, scientists are nowhere near managing to achieve this staggering feat of engineering, although admittedly technology is getting smaller all the time, and some believe that it is only a matter of time before such tiny machines are possible.

To give some idea of the scale we are talking about: one of the smallest objects visible to the naked eye is a grain of salt, which is approximately half a millimetre, or 500 microns (500 millionths of a metre) in size. The cell body of an average animal cell is much smaller (between 30 to 10 microns in diameter); and yet this is huge compared to say a virus which in some cases may be as small as 24 nanometres in length (a nanonmetre is a thousand millionth of a metre). Even then, viruses are enormous compared to atoms which may be a tenth of a nanometre across. The important question is whether we can really get down far enough to engineer at the atomic level. As early as 1959, in order to spur scientists into action, Richard Feynmann offered $1000 to anyone who could make a rotating electronic motor that was 64th of an inch in size (approximately the size of a full stop at the end of this sentence); or to take a page of writing that could be written (and read) on a head of a pin. By 1960 the prize for the motor had been won, although it took another 15 years before the second prize awarded. Today, nanoparticles are being created that are increasingly getting smaller, with well defined shapes and sizes, including 60 atom carbon 'buckyballs', and nanotubes that are only a few nanometres in diameter. With such miniature components, it may only be a matter of time before the first nanomachines are created.

It is impossible to tell whether nanotechnology is ever going to have an impact on aging or increasing longevity. In his book *The Engines of Creation*, Eric Drexler has argued that one day it may be possible to invent cell repair machines which are able to travel around the body correcting any damage that they come across. For example, they may be able to clean out our blood vessels, repair cells that have accumulated free radical damage, and replace cells that have been destroyed. In fact, if Drexler's dream ever came true, and the body with all its trillions of cells could be repaired at the particle (or even molecular level), then it is probable that aging would be reversible, and immortality ours to keep (in theory, we could even be involved in a fatal accident and like Humpty Dumpty be put back together again). However, it is probably fair to say that most scientists believe that such a prospect is, and will forever be, impossible. But, there again, how else does one undo the process of aging?

CRYONICS

The immediate freezing at very low temperatures of the body following death in the hope of avoiding decomposition, and then being thawed and revived later on, would appear to be our last resort of obtaining immortality. There are three cryonic organisations in the United States, the largest being the Alcor Life Extension Foundation, a non profit foundation based in California, with branches in Australia and Britain. People who choose this route to immortality can either have their whole body preserved at the cost of $120,000 or have the cheaper "neuro" option in which the head is preserved at the cost of around $40,000 (1992 prices). Either way, as soon as possible after the person has been confirmed as dead, the body is prepared by Alcon technicians for storage (an ambulance is always waiting to collect the body). To begin, the body is injected with anti-clotting agents and connected to a heart-lung machine which attempts to keep the blood circulating through the body in order to maintain it with a supply of oxygen and nutrients (this is particularly important to minimise damage to the brain). The blood is then replaced with an antifreeze solution that cools the corpse down to around 2°C, and then over the next 2 days the body is gradually cooled further to a temperature of −79°C (during this time it is also injected with cryoprotective drugs). The body is then placed in to a special bag and placed into a capsule of liquid nitrogen at −196°C, at which point the body is in a state of suspension where all biological decay has ceased.

The hope, of course, is that the body will be preserved so that it can be brought back to life at a future date. However, despite these valiant attempts to preserve the body, it is clear that the procedure must, nevertheless, cause considerable damage to its tissues. Although it is possible to freeze some simple tissues that only contain a few cells successfully, such as sperm, white blood cells and young embryos (indeed, there are people who began life by being frozen in liquid nitrogen in this way), cryopreserving complex tissues is much more difficult. For example, all tissues contain water (even cryopreserved bodies will contain a small amount) and when frozen it expands bursting small blood vessels, distorting structure, and destroying delicate cellular membranes. In addition, it must be the case that whole tissue cannot be viably preserved unless the cryoprotective agent evenly penetrates *every* cell which is unlikely to occur. Moreover, the many different types of cell in the body probably require different rates of freezing (and thawing) to best maintain their viability. Finally, the most important organ of all for those hoping to return from their suspension, namely the brain, also happens to be the most delicate and complex of bodily structures. Indeed, the brain shows considerable damage when frozen and thawed, and it is hard to imagine that this does not also occur when the brain is stored at $-196°C$ in liquid nitrogen! And, on top of all this, the patient is hoping that one day they will be thawed out, any damage that has occurred to their body put right (including the fatal disease or cause of death), and the body finally re-animated.

As Leonard Hayflick has pointed out, modern day cryonists might be compared with the ancient Egyptians who believed that mummification would allow the deceased to carry on in the afterlife. Indeed, most biologists would undoubtedly give the present cryogenic practice about the same chance of resurrecting life as that of mummification (perhaps even less, because it might be a more viable option of eventually reconstituting someone from a sample of their DNA as suggested in the film Jurassic Park). In fact, considering the cost of the cryonic procedure, one might be better advised to spend their money on making the most of this life, rather than the slim chance of surviving to the next. But, if one has the money to spare, then it can also be argued that one has nothing to lose, and even the barest glimmer of hope may be more reassuring than no hope at all. Despite this, it is hard to disagree with Morton Schatzman who has pointed out in his 1992 *New Scientist* article that all the evidence indicates – at least for the present – that it is simply an unusual and expensive way of disposing of dead bodies.

THE NEED FOR A UNIFIED THEORY OF AGING

So far in this chapter we have discussed a number of ways in which future developments may play a role in helping us to understand, and to control the aging process. But perhaps the most important development of all will occur when scientists are able to understand with confidence the causes of aging, and to establish how these change the fabric of the body to produce its biological decline. Arguably, it is only when scientists have this knowledge that they will be able to realistically work towards finding ways of slowing down or overcoming the aging decline – whether it is through genetic engineering or by the development of anti-aging drugs. Unfortunately, as this book has shown, we still have no adequate theory of biological aging; and even more depressingly, there are several fundamental questions that need to be solved before any real progress can be made. However, despite its great complexity, there is no reason why science cannot explain the process of aging, from cause to effect, in all its entirety. This must surely be a realistic aim for the foreseeable future, and one that researchers should be striving for.

Despite this, it is difficult to see what form such a theory will take. A successful theory of aging is likely to be hugely complex combining an interactive multitude of genetic and environmental causes. It may be that different causes of aging are occurring at the genetic, cellular and organ levels, and all are combining to produce biological decline (although it could turn out to be more simple than this). Furthermore, it may be that as more factors are found to be involved in aging – it will be realised that we all age in our own unique and individual way – making an unified theory even more difficult to formulate. In short, a successful theory of aging is likely to be one of the most complex (if not the most complex) that biology can ever hope to develop. Indeed, trying to understand aging and encapsulate it in a theory is a daunting prospect, and some doubt that it can be done. However, it remains the central key to unlocking the puzzle, and we have little option but to go forward, and hopefully one day spring the lock. Hopefully, as this book has shown, we have at least the foundations for a solution.

FUTURE PROSPECTS

What then can we expect in the future? Clearly, it is impossible to predict the future and any attempt to do so is bound to appear ludicrous within a few years. Indeed, anyone who thinks otherwise should bear in mind the words of Arthur C. Clarke:

> *with monotonous regularity, competent men have laid down the law about what is technically possible or impossible – and have been proved utterly wrong, sometimes while the ink was scarely dry from their pens.*

Furthermore, there are many amusing examples that can be used to illustrate Clarke's quote. For example, before locomotives were built there were experts who confidently asserted that suffocation would occur if people travelled over 30 miles an hour. Similarly, around the turn of the century, nearly all scientists were unanimous (and indeed often vociferous in their arguments) that heavier-than-air flight was impossible. Moreover, examples of pessimists who have been proven wrong continue to the present day. For example, in 1971, Sir Frank Macfarlane Burnet (who shared the 1945 Nobel Prize for physiology and medicine) commenting about the possibility of treating illness by inserting DNA into cells (i.e., gene therapy) wrote that he should be willing to state in any company that the chance of doing this type of work would remain infinitely small to the last syllable of time. And, of course, as we have seen above, within 20 years of this claim, Burnet's predictions were proven completely and utterly wrong.

If there is one thing to be learned from this, it is that acknowledged experts in the field are generally the ones who provide the most unreliable pointers to the future. In fact, Arthur C. Clarke has even gone so far to say that history shows that when a scientist states that something is possible, he is almost certainly right; and when he states that something is impossible, he is often proved wrong. Expert knowledge it would seem is not conducive to prophecy. Much more important is having the imagination to believe that something will one day become possible. On this basis, therefore, it would seem that the pessimists who argue that aging is too complex to be understood, or too intractable to be overcome, are indeed the ones who will one day be most likely to be proven wrong. But, alternatively, it cannot be totally ruled out that aging may turn out to be the one exception that bucks the trend. Indeed, in support of this notion is the fact that there have been many forecasts about aging made in the past that have been made to look ridiculous with the test of time. For example, Bernard Strehler, one of the most respected gerontologists of his time, and author of the 1962 book *Time, Cells and Aging*, predicted in 1971 that we would understand the aging process within 5 to 10 years, and be able to restore function in 10 to 30 years. Perhaps, more to the point, Strehler was not alone in his views, and reading literature from the 1960s and 1970s one can find many other examples of similar predictions. Thus, up to now, the pessimists have been proven right.

It would be fascinating to survey the attitudes of current leading researchers working in the biology of aging to find out whether they believe that aging will be understood, or conquered, in the coming years. However, it does seem that in general, there is considerable optimism about future prospects. For example, Michael Rose at the University of California has stated:

If we are willing to spend enough money on research, and if we spend it intelligently, I believe that in 25 years we could see the creation of the first products that can postpone human aging significantly. This would be only the beginning of a long process of technological development in which human life span would be aggressively extended. The only practical limit to human life span is the limit of human technology.

And, in a similar, if not more optimistic vein, Edward Masoro at the University of Texas has been quoted as saying:

Once we understand the mechanisms that control aging, we may find it possible to extend life span considerably more, perhaps by 100% – which would give us an extra 100 to 120 years.

These views do not seem to be untypical of researchers working in the field (e.g., see Brad Darrach's 1992 article 'The War on Aging' in *Life Magazine*) although admittedly there are many differences of opinion. In short, there appears to be general optimism that aging will one day be understood, and following this, the development of technology that will allow scientists to intervene in the process. Nevertheless, one must remember that understanding the aging process is not the same thing as having the technology to control it. For example, we understand the cause of many genetic diseases but we still await the technology to exploit this knowledge. But, at least understanding shows the way forward and shines a light on what was once dark. Unfortunately, with aging, we still await the basic understanding breakthrough. However, when is this likely to occur? One estimate has come Caleb Finch, author of *Longevity, Senescence and the Genome*, who has been quoted as saying (in 1992) that over the next 20 to 30 years, the study of aging mechanisms will create a crowning intellectual synthesis of currently separate biomedical approaches to the problem. In short, if Finch is right then we should understand all the complexities of aging by 2022. The task will then be to develop the technologies to intervene in the process.

But, of course, the real $64 million dollar question is: when are we going to be able to do this and to achieve immortality? Perhaps we should go back to Arthur C. Clarke who has a better record than most in the art of predicting the future. For example, back in 1945 he predicted that communication satellites would orbit the earth, and in 1947 he correctly predicted 1959 as the year in which the first moon rocket would be launched. He also predicted the first manned Apollo mission to reach the moon – although he was 9 years out by predicting this event in 1978. What then about aging and immortality? In his book *Profiles of the Future* written in 1962, Clarke predicted a number of scientific accomplishments and estimated that humans would achieve immortality around the year 2095. This is, in approximately 100 years time. Is this far-fetched? It seems not. Indeed, the general consensus seems to be, that sooner or later, it can be done – after all, aging is simply a biological problem. And when that day finally arrives, it is certain that the world will be changed for ever.

SUMMARY

- It has been estimated that the amount of scientific knowledge is doubling every 10 years. This means that we should have accumulated twice as much knowledge about aging in 10 years time, and 4 times as much in 20 years.
- The aim of the Human Genome Project is to determine the complete sequence of the 3 thousand million bases that make up our genome. One of the first benefits of this quest will be in the detection of genetic disease; but in the long term it should also help establish the function of every gene in the body.
- Research extending from the Human Genome Project may one day help identify all the genes involved in aging and solve the question of whether aging is a programed or stochastic process.
- Genetic engineering refers to a group of techniques that involves artificially altering the genetic make-up of an organism and usually involves transferring the DNA of one organism into another.
- It may be possible one day to genetically engineer higher organisms (including humans) to live longer. Such work has already been undertaken with fruit flies.
- Lifestyle factors are probably just as important (if not more so) in determining human longevity. Moreover, wear and tear, particularly at the molecular level (including free radical damage) is likely to be a very important cause of aging. It is difficult to see how genetic engineering can ever protect fully against such processes.
- Nanotechnology holds the promise that one day it may be possible to undo, or reverse, the damage done by aging. However, most scientists believe that such claims will never become feasible.

- Cryonics offers the hope that the dead corpse can be preserved until a way can be found in the future to bring it back to life. However, it is clear that the cryonic embalming causes cellular damage, particularly to the brain, and it is difficult to see how such damage could ever be successfully corrected.
- Perhaps the most important advance in the fight against aging will be the development of a unified theory that can explain all the important causes of aging, their interactions, and how they act on each level of biological organisation to produce biological decline. This should enable researchers to focus more effectively on the best ways to stop aging.
- There are grounds for believing that a successful theory of aging will be developed by the year 2020, and if Arthur C. Clarke is right, then the technology required to produce immortality in humans should be in place by the year 2095.

REFERENCES

GENERAL

Adelman, R.C. and Roth, G.S. eds (1983) *Testing the Theories of Aging.* Boca Raton: CRC Press.

Arking, R. (1991) *Biology of Aging: Observations and Principles.* Englewood Cliffs: Prentice Hall.

Austad, S.N. (1997) *Why We Age: What Science is Discovering About the Body's Journey Through Life.* New York: John Wiley.

Bellamy, D. (1995) *Aging: A Biomedical Perspective.* Chichester: John Wiley.

Barash, D.P. (1983) *Aging: An Exploration.* Seattle: University of Washington Press.

Behnke, J.A., Finch, C.E. and Moment. G.B. eds (1979) *The Biology of Aging.* New York: Plenum Press.

Comfort, A. (1979) *The Biology of Senescence.* New York: Elsevier.

Davies, I. (1983) *Ageing.* London: Edward Arnold.

DiGiovanna, A.G. (1994) *Human Aging: Biological Perspectives.* New York: McGraw-Hill.

Finch, C.E. (1990) *Longevity, Senescence, and the Genome.* Chicago: Uiversity of Chicago Press.

Finch, C.E. and Hayflick, L. eds (1977) *Handbook of the Biology of Aging.* New York: Van Nostrand.

Finch, C.E. and Schneider, E.L. (1985) *Handbook of the Biology of Aging,* 2nd edn. New York: Van Nostrand.

Freeman, J.T. (1979) *Aging: Its History and Literature.* New York: Human Sciences Press.

Goldstein, S. (1993) The biology of aging: looking to defuse the genetic time bomb. *Geriatrics,* **48,** 76–82.

Gosden, R. (1996) *Cheating Time: Science, Sex and Aging.* London: Macmillan.

Hart, R.W. and Turturro, A. (1983) Theories of aging. In *Review of Biological Research in Aging,* edited by Rothstein, M., pp 5–17. New York, Alan R. Liss.

Hayflick, L. (1985) Theories of biological aging. *Experimental Gerontology,* **20,** 145–159.

Hayflick, L. (1996) *How and Why We Age.* New York: Ballantine Books.

Holliday, R. (1995) *Understanding Ageing*. Cambridge: Cambridge University Press.

Kahn, C. (1985) *Beyond the Helix*. New York: Time Books.

Lamb, M.J. (1977) *Biology of Aging*. London: Blackie.

Maddox, G.L. ed (1995) *The Encyclopedia of Aging*. New York: Springer.

Medina, J.J. (1996) *The Clock of Ages*. Cambridge: Cambridge University Press.

Prehoda, R.W. (1968) *Extended Youth*. New York: G.P. Putnam.

Ricklefs, R.E. and Finch, C.E. (1995) *Aging: A Natural History*. New York: Scientific American Library.

Rosenfeld, A. (1976) *Prolongevity*. New York: Alfred A. Knopf.

Schofield, J.D. and Davies, I. (1978) Theories of aging. In *Textbook of Geriatric Medicine and Gerontology*, edited by Brocklehurst, J.C., pp 37–70. Edinburgh: Churchill Livingstone.

Schneider, E.L. and Rowe, J.W. (1990) *Handbook of the Biology of Aging*, 3rd edn. San Diego: Academic Press.

Strehler, B.L. (1982) *Time, Cells and Aging*. New York: Academic Press.

Walford, R.L. (1984) *Maximun Life Span*. New York: W.W. Norton.

Warner, H.R. *et al* (1987) *Modern Biological Theories of Aging*. New York: Raven Press.

CHAPTER 1: THE CHARACTERISTICS OF AGING

Barinaga, M. (1991) How long is the human life span? *Science*, **254**, 936–938.

Blumenthal, H.T. (1978) Aging: biologic or pathologic? *Hospital Practice*, April, 127–137.

Carey, J.R. *et al* (1992) Slowing of mortality at older ages in large medfly cohorts. *Science*, **258**, 457–461.

Central Statistical Office. (1993) *Social Trends*, **23**, chap 7. London: HMSO.

Craig, J. (1994) Centenarians: 1991 estimates. *Population Trends*, **75**, 30–32.

Dura, R. *et al* (1984) Human brain glucose utilization and cognitive function in relation to age. *Annals of Neurology*, **16**, 702–713.

Fries, J.F. (1980) Aging, natural death, and the compression of morbidity. *New England Journal of Medicine*, **303**, 130–135.

Fries, J.F. and Crappo, L.M. (1981) *Vitality and Aging.* New York: Freeman and Co.

Fries, J.F. (1984) The compression of Mortlaity: miscellaneous comments about a theme. *The Gerontologist,* **24**, 354–359.

Gompertz, B. (1825) On the nature of the function of the laws of human mortality. *Philosophical Transections of the Royal Society of London,* **1**, 513–585.

Hadley, E.C. (1992) Causes of death among the oldest old. In *The Oldest Old,* edited by Suzman, R.M., Willis, D.P. and Manton, K.G., pp 183–196. Oxford: Oxford University Press.

Harrison, D.E. (1982) Must we grow old? *Biology Digest,* **8**, 11–25.

Herbert, S. (1987) The oldest person in the world dies at 122. *The Daily Telegraph.* August 5th.

Kohn, R.R. (1982) Causes of death in very old people. *JAMA,* **247**, 2793–2797.

Krag, C.L. and Kountz, W.B. (1950) Stability of body function in the aged: effect of exposure of the body to cold. *Journal of Gerontology,* **5**, 227–235.

Lakatta, E.G. (1987) Human aging: changes in structure and function. *Journal of the American College of Cardiology,* **10**, 42A–47A.

Makeham, W.H. (1867) On the law of mortality. *Journal of Insurance Actuaries,* **13**, 325–358.

Masoro, E.J. (1991) Biology of aging: facts, thoughts and experimental approaches. *Laboratory Investigation,* **65**, 500–510.

Medvedev, Z.A. (1990) An attempt at a rational classification of theories of aging. *Biological Research,* **65**, 375–398.

Nuland, S.B. (1993) *How We Die.* London: Chatto and Windrus.

Office of Population Censuses and Surveys. (1993) *Mortality Statistics: General, England and Wales,* DH1 no27. London: OPCS.

Olshansky, S.J. *et al* (1990) In search of Methuselah: estimating the upper limits to human longevity. *Science,* **250**, 634–640.

Sapolsky, R. and Finch, C.E. (1991) On growing old. *The Sciences,* **30**, 30–38.

Schneider, E.L. and Brody, J.A. (1983) Aging, natural death and the compression of mortality: another view. *New England Journal of Medicine,* **309**, 854–856.

Shock, N.W. (1977) Systems integration. In *Handbook of the Biology of Aging,* edited by Finch, C.E. and Hayflick, L., pp 639–665. New York, Van Nostrand.

Spirduso, W.W. (1980) Physical fitness and psychomotor speed: a review. *Journal of Gerontology,* **35**, 850–865.

CHAPTER 2: THE EVOLUTION OF AGING

Casti, J.L. (1989) *Paradigms Lost: Images of Man in the Mirror of Science.* London: Sphere Books.

Cutler, R.G. (1979) Evolutionary biology of senescence. In *The Biology of Aging*, edited by Behnke, J.A., Finch, C.E. and Moment, G.B. eds, pp 311–360. New York: Plenum Press.

Cutler, R.G. (1991) Antioxidants and aging. *American Journal of Clinical Nutrition*, **53**, 373S–379S.

Dawkins, R. (1976) *The Selfish Gene.* Oxford: Oxford University Press.

DeDuve, C. (1996) The birth of complex cells. *Scientific American*, April, 38–45.

Kent, S. (1980) The evolution of longevity. *Geriatrics*, Jan, 98–104.

Kirkwood, T.B.L. (1977) Evolution of aging. *Nature*, **270**, 301–304.

Kirkwood, T.B.L. (1985) Comparative and evolutionary aspects of longevity. In *Handbook of the Biology of Aging*, 2nd edn, edited by Finch, C.E. and Schneider, E.L. pp 27–44, New York: Van Nostrand.

Kirkwood, T.B.L. (1992) Comparative life spans of species: why do species have the life spans they do? *American Journal of Clinical Nutrition*, **55**, 1191S–1195S.

Medawar, P.B. (1952) *An Unsolved Problem in Biology.* London: H.K. Lewis.

Milner, R. (1990) *The Encyclopedia of Evolution.* New York: Facts on File.

Rose, M.R. (1991) *Evolutionary Biology of Aging.* Oxford: Oxford University Press.

Rose, M.R. and Charlesworth, B. (1980) A test of evolutionary theories of senescence. *Nature*, **287**, 141–142.

Sacher, G.A. (1968) Molecular versus systemic theories on the genesis of aging. *Experimental Gerontology*, **3**, 265–271.

Sacher, G.A. (1978) Longevity and aging in vertebrate evolution. *BioScience*, **28**, 497–501.

Sacher, G.A. (1982) Evolutionary theory in gerontology. *Perspectives in Biology and Medicine*, **25**, 339–353.

Sonneborn, T.M. (1979) The origin, evolution, nature, and causes of aging. In *The Biology of Aging*, edited by Behnke, J.A., Finch, C.E. and Moment, G.B. eds, pp 361–375. New York: Plenum Press.

Stearns, S.C. (1992) *The Evolution of Life Histories.* Oxford: Oxford University Press.

Weismann, A. (1889) *Essays Upon Heredity and Kindred Biological Problems.* Oxford: Clarendon Press.

Williams, G.C. (1957) Pleiotrophy, natural selection, and the evolution of senescence. *Evolution*, **11**, 398–411.

CHAPTER 3: GENES AND AGING

Abbott, M.H. *et al* (1974) The familial component of longevity. A study of offspring on nonagenarians. *John Hopkins Medical Journal*, **134**, 1–16.

Bell, A.G. (1918) *The Duration of Life and Conditions Associtaed with Longevity*. Washington D.C.: Genealogical Research Office.

Bouchard, T.J. *et al* (1990) Sources of human psychological differences: the Minnesota study of twins reared apart. *Science*, **225**, 223–228.

Brown, W.T. (1987) Genetic aspects of aging in humans. *Review of Biological Research in Aging*, **3**, 77–91.

Cohen, B.H. (1964) Family patterns of mortality and life span. *Quarterly Review of Biology*, **39**, 130–181.

Concur, D. (1996) Mutants tear up nature's timetable. *New Scientist*. May. p 16.

DeBusk, F.L. (1972) The Hutchinson–Gilford progeria syndrome. *Journal of Pediatrics*, **80**, 697–724.

Dublin, L.I. (1949) *Length of Life*. New York: The Ronald Press.

Dudas, S.P. and Arking, R. (1995) A Coordinate upregulation of anti-oxidant gene activities is asscoited with delayed onset of senescence in a long-lived strain of Drosophilia. *Journal of Gerontology: Biological Sciences*, **50**, 117–127.

Goto, M. *et al* (1992) Genetic linkage of Werner's syndrome to five markers on chromosome 8. *Nature*, **355**, 735–738.

Holden, C. (1980) Identical twins reared apart. *Science*, **207**, 1323–1327.

Holden, C. (1987) Why do women live longer than men? *Science*, **238**, 158–160.

Johnson, T.E. (1990) Increased life span of age-1 mutants in Caenorhabditis elegans and lower Gompertz rate of aging. *Science*, **249**, 908–912.

Kallman, F.J. (1957) Twin data on the genetics of aging. In *Methodology of the Study of Aging*, edited by Wolstenholme, G.E.W. and O'Connor, C.M. pp 131–148. Boston: Little Brown and Co.

Lakowski, B. and Hekimi, S. (1996) Determination of life span in Caenorhabditis elegans by four clock genes. *Science*, **272**, 1010–1013.

Luckinbill, L.S. *et al* (1988) Localising genes that defer senescence in Drosophila melanogaster. *Heredity*, **60**, 367–374.

Martin, G.M. (1978) Genetic syndromes in man with potential relevance to the pathobiology of aging. In *Genetic Effects on Aging* edited by Bergsma, D. and Harrison, D.E., pp 5–40. New York: Alan R. Liss.

Orr, W.C. and Sohal, R.S. (1994) Extension of life span by overexpression of superoxide dismutase and catalase in Drosophila melanogaster. *Science*, **263**, 1128–1130.

Pearl, R. and Pearl, R. (1934) *The Ancestory of the Long-Lived*. Baltimore: John Hopkins Press.

Rose, M.R. (1984) Laboratory evolution of postponed senescence in Drosophilia melanogaster. *Evolution*, **38**, 1004–1010.

Rusting, R.L. (1992) Why do we age? *Scientific American*. Dec, 86–95.

Service, P.M. *et al* (1985) Resistance to environmental stress in Drosophila melanogaster selected for postponed senescence. *Physiological Zoology*, **58**, 380–389.

Smith, D.W.E. (1993) *Human Longevity*. Oxford: Oxford University Press.

Wood, W.B. and Johnson, T.E. (1994) Stopping the clock. *Current Biology*, **4**, 151–153.

CHAPTER 4: GENETIC THEORIES OF AGING

Clark, W.R. (1996) *Sex and the Origins of Death*. Oxford: Oxford University Press.

Cutler, R.G. (1973) Redundancy of information content in the genome of mammilian species as a protective mechanism determining aging rate. *Mechanisms of Aging and Development*, **2**, 381–408.

Duke, R.C., Ojcius, D.M. and Young, J.D. (1996) Cell suicide in health and disease. *Scientific American*. Dec. 48–55.

Fargnoli, J. *et al* (1990) Decreased expression of heat shock protein 70 mRNA and protein after heat treatment in cells of aged rats. *Proceedings of the National Academy of Sciences, USA*, **87**, 846–850.

Gensler, H.L. and Bernstein, H. (1981) DNA damage as the primary cause of aging. *The Quaterly Review of Biology*, **56**, 279–303.

Hall, D.A. (1978) The aging process: two theories explained. *Modern Geriatrics*. Sept. 60–64.

Hart, R. and Setlow, R.B. (1974) Correlation between deoxyribonucleic acid excision-repair and life span in a number of mammalian species. *Proceedings of the New York Academy of Sciences, USA*, **71**, 2169–2173.

Jones, S. (1993) *The Language of the Genes*. London: Flamingo.

Kanungo, M.S. (1994) *Genes and Aging*. Cambridge: Cambridge University Press.

Lints, F.A. (1978) Genetics and aging. *Interdisciplinary Topics in Gerontology*, **14**, 1–130.

Liu, A.Y. *et al* (1989) Attenuated induction of heat shock gene expression in aging diploid fibroblasts. *The Journal of Biological Chemistry*, **264**, 12037–12045.

Martin, G.M. (1981) Genotropic theories of Aging. In *Aging, Cancer and Cell Membranes* edited by Borek, C., Fenoglio, C.M. and King, D.W. pp 5–20, Stuttgart: Verlag.

Medvedev, Z.A. (1975) Aging and longevity: new approaches and new perspectives. *The Gerontologist*. June. 196–201.

Morrow, J. and Garner, C. (1979) An evaluation of some theories of the mechanism of aging. *Gerontology*, **25**, 136–144.

Russell, R.L. (1987) Evidence for and against the theory of developmentally programmed aging. In *Modern Biological Theories of Aging* edited by Warner, H.R. *et al* New York: Raven Press.

Thakur, M.K., Oka, T. and Natori, Y. (1993) Gene expression and aging. *Mechanisms of Aging and Development*, **66**, 283–298.

Tice, R.R and Setlow, R.B. (1985) DNA repair and replication in aging organisms and cells. In *Handbook of the Biology of Aging* edited by Finch, C.E. and Schneider, E.L. pp 173–224. New York: Van Nostrand.

CHAPTER 5: THE CELLULAR BASIS OF AGING

Adelman, R.C. and Dekker, E.E. eds (1985) *Modification of Proteins During Aging*. New York: Alan R. Liss.

Ames, B.N., Shigenaga, M.K. and Hagan, T.M. (1993) Oxidants, antioxidants, and the degenerative diseases of aging. *Proceedings of the National Academy of Science, USA*, **90**, 7915–7922.

Barrows, C.H. (1971) The challenge – mechanisms of biological aging. *The Gerontologist*, Spring, 5–11.

Bjorkstein, J. (1968) The crosslinkage theory of aging. *Journal of the American Geriatrics Society*, **16**, 408–427.

Carper, J. (1995) *Stop Ageing Now!* London: HarperCollins.

Enstrom, J.E. and Pauling, L. (1982) Mortality among health-conscious elderly Californians. *Proccedings of the National Academy of Science, USA*, **79**, 6023–6027.

Gershon, D. and Gershon, H. (1976) An evaluation of the "error catastrophe" theory of ageing in the light of recent experimental results. *Gerontology*, **22**, 212–219.

Harman, D. (1981) The aging process. *Proceedings of the National Academy of Sciences, USA*, **78**, 7124–7128.

Harman, D. (1984) Free radical theory of aging: the free radical diseases. *Age*, **7**, 111–131.

Meir, M.J. *et al* (1993) Vitamin E consumption and the risk of coronary disease in women. *New England Journal of Medicine*, **328**, 1444–1449.

Pearson, D. and Shaw, S.(1982) *Life Extension*. New York: Warner Books.

Rimm, E.B. *et al* (1993) Vitamin E consumption and the risk of coronary heart disease in men. *New England Journal of Medicine*, **328**, 1450–1456.

Rothstein, M. (1982) *Biochemical Approaches to Aging*. New York: Academic Press.

Stephens, N.G. *et al* (1996) Randomised controlled trial of vitamin E in patients with coronary disease. *The Lancet*, **347**, 781–786.

Tolmasoff, J.M. *et al* (1980) Superoxide dismutase: correlation with life-span and specific metabolic rate in primate species. *Proceedings of the National Academy of Sciences, USA*, **77**, 2777–2781.

Yu, B.P. ed (1993) *Free Radicals in Aging*. Boca Raton: CRC Press.

CHAPTER 6: THE HAYFLICK LIMIT

Bayreuther, K. *et al* (1988) Human skin fibroblasts *in vitro* differentiate along a terminal cell lineage. *Proceedings of the National Academy of Sciences, USA*, **85**, 5112–5116.

Bell, E. *et al* (1978) Loss of division *in vitro*: aging or differentiation? *Science*, **202**, 1158–1163.

Campisi, J. (1992) Oncogenes, proto-oncogenes and tumour suppressor genes: a hitchhikers guide to senescence? *Experimental Gerontology*, **27**, 397–402.

Cooper, G.M. (1992) *Elements of Human Cancer*. Boston: Jones and Bartlett.

Cristofalo, V.J. and Pignolo, R.J. (1993) Replicative senescence of human fibroblast-like cells in culture. *Physiological Reviews*, **78**, 617–638.

Dice, J.F. (1993) Cellular and molecular mechanisms of aging. *Physiological Reviews*, **73**, 149–159.

Goldstein, S. (1990) Replicative senescence: the human fibroblast comes of age. *Science*, **249**, 1129–1133.

Harley, C.B. *et al* (1990) Telomeres shorten during ageing of human fibroblasts. *Nature*, **345**, 458–460.

Hayflick, L. (1968) Human cells and aging. *Scientific American*, **218**, 32–37.

Hayflick, L. (1974) The strategy of senescence. *The Gerontologist*, February, 37–45.

Hayflick, L. (1980) The cell biology of huamn aging. *Scientific American*, **242**, 42–49.

Hayflick, L. (1992) Aging, longevity, and immortality *in vitro*. *Experimental Gerontology*, **27**, 363–368.

Koli, K. and Keski-Oja, J. (1992) Cellular senescence. *Annals of Medicine*, **24**, 313–318.

Lewis, C.M. and Tarrant, G.M. (1972) Error theory and ageing in human diploid fibroblasts. *Nature*, **239**, 316–318.

Pereira-Smith, O.M. (1992) Molecular genetic approaches to the study of celular aging. *Experimental Gerontology*, **27**, 441–445.

Phillips, P.D. and Cristofalo, V.J. (1987) A review of recent research on cellular aging in culture. *Review of Biological Research in Aging*, **3**, 385–415.

Rohme, D. (1981) Evidence for a relationship betwen longevity of mammalian species and life spans of normal fibroblasts *in vitro* and erythrocytes *in vivo*. *Proceedings of the National Academy of Sciences, USA*, **78**, 5009–5013.

Scientific American, Inc (1997) *What You Need to Know About Cancer*. New York: W.H. Freeman.

Smith, J.R. and Lincoln, D.W. (1984) Aging of cells in culture. *International Review of Cytology*, **89**, 151–177.

Steel, M. (1995) Telomerase that shapes our ends. *The Lancet*, **345**, 935–936.

West, M.D. *et al* (1989) Replicative senescence of human skin fibroblasts correlates with a loss of regulation and overexpression of collagenase activity. *Experimental Cell Research*, **184**, 138–147.

Witkowski, J.A. (1979) Alexis Carrel and the mysticism of tissue culture. *Medical History*, **23**, 279–296.

Witkowski, J.A. (1980) Dr.Carrel's immortal cells. *Medical History*, **24**, 129–142.

CHAPTER 7: METABOLISM AND AGING

Austad, S.N. and Fischer, K.E. (1991) Mammalian aging, metabolism and ecology. *Journal of Gerontology: Biological Sciences*, **46**, B47–53.

Baker, G.T. (1976) Insect flight muscle: maturation and senescence. *Gerontology*, **22**, 334–361.

Cerami, A. (1985) Glucose as a mediator of aging. *Journal of the American Geriatric Society*, **33**, 626–634.

Economos, A.C. (1981) Beyond the rate of living. *Gerontology*, **27**, 258–265.

Harman, D. (1972) The biologic clock: the mitochondria? *Journal of the American Geriatric Society*, **20**, 145–147.

Linnane, A.W. *et al* (1990) Mitochondrial gene mutation: the aging process and degenerative diseases. *Biochemistry International*, **22**, 1067–1076.

Lints, F.A. (1989) The rate of living theory revisited. *Gerontology*, **35**, 36–57.

Liu, R.K. and Walford, R.L. (1975) Mid-life temperature-transfer effects on life span of annual fish. *Journal of Gerontology*, **30**, 129–131.

Maynard Smith, J. (1962) Review lectures on senescence. 1. the causes of ageing. *Proceedings of the Royal Society of London. Series B*, **157**, 115–127.

McGandy, R.B. *et al* (1966) Nutrient intake and energy expenditure in men of different ages. *Journal of Gerontology*, **21**, 581–587.

Miquel, J. *et al* (1980) Mitochondrial role in cell aging. *Experimental Gerontology*, **15**, 575–591.

Miquel, J. (1991) An integrated theory of aging as the result of mitochondrial-DNA mutation in differentiated cells. *Archives of Gerontology and Geriatrics*, **12**, 99–117.

Rowe, J.W. and Troen, B.R. (1980) Sympathetic nervous system and aging in man. *Endocrine Reviews*, **1**, 167–179.

Sohal, R.S. (1976) Aging changes in insect flight muscle. *Gerontology*, **22**, 317–333.

Sohal, R.S. (1976) Metabolic rate and life span. *Interdisciplinary Topics in Gerontology*, **9**, 25–40.

Vaughan, L *et al* (1991) Aging and energy expenditure. *American Journal of Clinical Nutrition*, **53**, 821–825.

CHAPTER 8: HORMONES AND AGING

Andres R. and Tobin, J.D. (1977) Endocrine Systems. In *Handbook of the Biology of Aging*, edited by Finch, C.E. and Hayflick, L.H. pp 357–378, New York: Van Nostrand.

Arendt, J. (1996) Melatonin. *British Medical Journal*, **312**, 1242–1243.

Barrett-Connor, E. *et al* (1986) A prospective study of dehydroepiandrosterone sulfate, mortality, and cardiovascular disease. *New England Journal of Medicine*, **315**, 1519–1524.

Corpas, E. *et al* (1993) Human growth hormone and human aging. *Endocrine reviews*, **14**, 20–39.

Davis, P.J. (1979) Endocrinology and aging. In *The Biology of Aging*, edited by Behnke, J.A. *et al*. pp 263–276. New York: Plenum Press.

Denckla, W.D. (1974) Role of the pituitary and thyroid glands in the decline of minimal O_2 consumption with age. *The Journal of Clinical Investigation*, **53**, 572–581.

Everitt, A.V. (1966) The pituitary gland: relation to aging and diseases of old age. *Postgraduate Medicine*, Nov, 645–650.

Iranmanesh, A. *et al* (1991) Age and relative adiposity are specific negative determinants of the frequency and amplitude of growth hormone (GH) secretory bursts and the half-life of endogenous GH in healthy men. *Journal of Clinical Endocrinology and Metabolism*, **73**, 1081–1088.

Korenmn, S.G. ed (1982) *Endocrine Aspects of Aging*. New York: Elsevier.

Lesnikov, V.A. and Pierpaoli, W. (1994) Pineal cross-transplantation (old to young and vice versa) as evidence for an endogenous "aging clock". *Annals of the New York Academy of Sciences*, **719**, 456–460.

Maestroni, G.J.M. *et al* (1988) Pineal melatonin, its fundamental immunoregulatory role in aging and cancer. *Annals of the New York Academy of Sciences*, **521**, 140–148.

Meites, J. ed (1983) *Neuroendocrinology of aging*. New York: Plenum Press.

Minaker, K.L. *et al* (1985) Endocrine systems. In *Handbook of the Biology of Aging. Second Edition*, edited by Finch, C.E. and Schneider, E.L. pp 433–456, New York: Van Nostrand.

Pierpaoli, W. *et al* (1991) The pineal control of aging: the effects of melatonin and pineal gland on the survival of older mice. *Annals of the New York Academy of Sciences*, **621**, 291–313.

Pierpaoli, W. and Regelson, W. (1994) Pineal control of aging: effect of melatonin and pineal grafting on aging mice. *Proceedings of the National Academy of Science*, **91**, 787–791.

Pierpaoli, W. and Regelson, W. (1995) *The Melatonin Miracle*. London: Fourth Estate.

Reppert, S.M. and Weaver, D.R. (1995) Melatonin madness. *Cell*, **83**, 1059–1062.

Rudman, D. *et al* (1990) Effects of human growth hormone in men over 60 years old. *New England Journal of Medicine*, **323**, 2–5.

Sapolsky, R.M. (1990) The adrenocotical axis. In *Handbook of the Biology of Aging. Third Edition*, edited by Schneider, E.L. and Rowe, J.W. pp 330–348, San Diego: New York.

Weiner, M.F. *et al* (1987) Influence of age and relative weight on cortisol suppression in normal subjects. *American Journal of Psychiatry*, **144**, 646–649.

Weksler, M.E. (1966) Hormone replacement for men. *British Medical Journal*, **312**, 859–860.

CHAPTER 9: SEX AND AGING

Aminoff, M.J. (1993) *Brown-Sequard: A Visionary of Science.* New York: Raven Press.

Borell, M. (1976) Organotherapy, british physiology, and the discovery of the internal secretions. *Journal of the History of Biology,* 9, 235–268.

Borell, M. (1985) Oranotherapy and the emergence of reproductive endocrinology. *Journal of the History of Biology,* 18, 1–30.

Drori, D. and Folman,Y. (1969) The effect of mating on the longevity of male rats. *Experimental Gerontology,* 4, 263–266.

Grady, D. *et al* (1992) Hormone therapy to prevent disease and prolong life in postmenopausal women. *Annals of Internal Medicine,* 117, 1016–1037.

Grodstein, F. *et al* (1996) Postmenopausal estrogen and progestin use and the risk of cardiovascular disease. *The New England Journal of Medicine,* 335, 453–461.

Hamilton. D. (1987) *The Monkey Gland Affair.* London: Chatto and Windus.

Hamilton, J.B. and Mestler, G.E. (1969) Mortality and survival: comparison of eunuchs with intact men and women in a mentally retarded population. *Journal of Gerontology,* 24, 395–411.

Henderson, B.E. *et al* (1991) Decreased mortality in users of estrogen replacement therapy. *Archives of International Medicine,* 151, 75–78.

Heriot, A. (1955) *The Castrati in Opera.* London: Secker and Warburg.

Kaler, L.W. and Neaves, W.B. (1978) Attrition of the human Leydig cell population with advancing age. *Anatomical Record,* 192, 513–518.

Medvei, V.C. (1993) *The History of Clinical Endocrinology.* Carnforth: Partenon Publishing.

Neaves, W.B. *et al* (1984) Leydig cell numbers, daily sperm production, and serum gonadotropin levels in aging men. *Journal of Clinical Endocrinology and Metabolism,* 59, 756–763.

Tenover, J.S. (1992) Effect of testosterone supplementation in the aging male. *Journal of Clinical Endocrinology and Metabolism,* 75, 1092–1098.

Van der Schouw, *et al* (1996) Age at menopause as a risk factor for cardiovascular mortality. *The Lancet,* 347, 714–718.

CHAPTER 10: THE IMMUNOLOGICAL BASIS OF AGING

Abbas, A.K. *et al* (1994) *Cellular and Molecular Immunolgy.* Philadelphia: W.B. Saunders.

Burnet, F.M. (1970) An immunological approach to aging. *The Lancet*, ii, 358–360.

Burnet, F.M. (1974) *Intrinsic Mutagenesis*. Leitchworth: Medical and Technical Publishing Co.

Clark, W.R. (1995) *At War Within*. New York: Oxford University Press.

Fabris, N. *et al* (1972) Lymphocytes, hormones and aging. *Nature*, **240**, 557–559.

George, A.J.T. and Ritter, M.A. (1996) Thymic involution with aging: obsolescence or good housekeeping? *Immunology Today*, **17**, 267–272.

Heidrick, M.L. and Makinodan, T. (1972) Nature of cellular deficiencies in age-related decline of the immune system. *Gerontologia*, **18**, 305–320.

Hooper, B. *et al* (1972) Autoimmunity in a rural community. *Clinical and Experimental Immunology*, **12**, 79–87.

Kay, M.M.B (1979) The thymus: clock for immunologic aging? *The Journal of Investigative Dermatology*, **73**, 29–38.

Mackay, I.R. (1972) Aging and immunological function in man. *Gerontologia*, **18**, 285-304.

Makinodan, T. and Kay, M.M.B. (1980) Age influence on the immune system. *Advances in Immunology*, **29**, 287–329.

Pawelec, G. *et al* (1995) Immunosenescence: aging of the immune system. *Immunolgy Today*, **16**, 420–422.

Roberts-Thomas, I.C. *et al* (1974) Ageing, immune response, and mortality. *The Lancet*, ii, 368–370.

Smith, G.S. and Walford, R.L (1977) Influence of the main histocompatibility complex on aging in mice. *Nature*, **270**, 727–729.

Takata, H. *et al* (1987) Influence of major histocompatibility complex region genes on human longevity among Okinawan-Japanese centenarians and nonagearians. *The Lancet*, ii, 824–826.

Walford, R.L. (1969) *The Immunologic Theory of Aging*. Copenhagen: Munksgaard.

Walford, R.L. (1974) Immunlogic theory of aging: current status. *Federation Proceedings*, **33**, 2020–2027.

Walford, R.L. (1982) Studies in immunogerontology. *Journal of the American Geriatrics Society*, **30**, 617–625.

Weksler, M.E. (1980) The immune system and the aging process in man. *Proceedings of the Society for Experimental Biology and Medicine*, **165**, 200–205.

Weksler, M.E. (1981) The senescence of the immune system. *Hospital Practice*, Oct, 53–64.

CHAPTER 11: CALORIE RESTRICTION AND LIFE EXTENSION

Carlson, A.J. and Hoelzel, F. (1946) Apparent prolongation of the life span of rats by intermittant feeding. *Journal of Nutrition*, **31**, 363–375.

Everitt, A.V. *et al* (1980) The effects of hypophysetomy and continous food restriction, begun at ages 70 and 400 days, on collagen aging. proteinura, incidence of pathology and longevity in the male rat. *Mechanisms of Aging and Development*, **12**, 161–172.

Leaf, A. (1973) Unusual longevity: the common denominators. *Hospital Practice*, Oct. 75–86.

Leaf, A. (1973) "Every day is a gift when you are over 100". *National Geographic*, **143**, 93–110.

Holehan, A.M. and Merry, B.J. (1986) The experimental manipulation of ageing by diet. *Biological Review*, **61**, 329–368.

Manson, J.E. *et al* (1987) Body weight and longevity: A reassessment. JAMA, 237, 353.

Masoro, E.J. *et al* (1982) Action of food restriction in delaying the aging process. *Proccedings of the National Academy of Sciences*, **79**, 4239–4241.

Masoro, E.J. *et al* (1991) Retardation of the aging process in rats by food restriction. *Annals of the New York Academy of Sciences*, **621**, 337–352.

McCarter, R. *et al* (1985) Does food restriction retard aging by reducing the metabolic rate? *American Journal of Physiology*, **248**, E488–E490.

McCay, C.M. *et al* (1935) Prolonging the life span. *Science Monthly*, **39**, 405–414.

Ross, M.H. (1961) Length of life and nutrition in the rat. *Journal of Nutrition*, **75**, 197–210.

Ross, M.H. (1972) Length of life and caloric intake. *American Journal of Clinical Nutrition*, **25**, 834–838.

Sorlie, P. *et al* (1980) Body build and mortality: the Framingham study. *Journal of the American Medical Association*, **243**, 1828–1831.

Tucker, M.J. (1979) The effects of long-term food restriction on tumours in rodents. *International Journal of Cancer*, **23**, 803–807.

Walford, R.L. (1985) The extension of life span. *Clinics in Geriatric Medicine*, **1**, 29–35.

Weindruch, R. (1996) Caloric restriction and aging. *Scientific American*, Jan, 32–38.

Weindruch, R. and Walford, R.L. (1988) *The Retardation of Aging and Disease by Dietary Restriction*, Springfield: Charles C. Thomas.

Yu, B.P. ed (1994) *Modulation of Aging Processes by Dietary Restriction.* Boca Raton: CRC Press.

Yu, B.P. (1994) How diet influences the aging process of the rat. *Proccedings of the Society for Experimental Biology and Medicine,* **205**, 97–105.

CHAPTER 12: FUTURE PROSPECTS

Anderson, W.F. (1984) Prospects for human gene therapy. *Science,* **226**. 401–409.

Brown, P. (1994) The new human condition. *New Scientist,* Oct, 8–11.

Clark, A.C. (1962) *Profiles of the Future.* London: Victor Gollancz.

Cherfas, J. (1982) *Man Made Life.* Oxford: Blackwells.

Concar, D. (1996) Death of old age. *New Scientist,* June, 24–29.

Cooper, N.G. ed (1994) *The Human Genome Project.* Mill Vally: University Science Books.

Darrach, B. (1992) The war on aging. *Life,* **15**, 33–43.

Drexler, K.E. (1986) *Engines of Creation.* New York: Doubleday.

Prehoda, R.W. (1968) *Extended Youth.* New York: Putman and Sons.

Rose, M.R. and Nusbaum, T.J. (1994) Prospects for postponing human aging. *The FASEB Journal,* **8**, 925–928.

Russo, E. and Cove, D. (1995) *Genetic Engineering: Dreams and Prospects.* Oxford: W.H. Frreman and Co.

Schatzman, M. (1992) Cold comfort at death's door. *New Scientist,* Sept, 36–39.

Segerberg, O. (1974) *The Immortality Factor.* New York: Dutton and Co.

Verma, I.M. (1990) Gene Therapy. *Scientific American,* Nov, 68–84.

Weatherall, D. (1995) *Science and the Quiet Art.* Oxford: Oxford University Press.

Wilkie,T. (1993) *Perilous Knowledge: The Human Genome Project and its Implications.* London: Faber and Faber.

INDEX

321